Light Engineering für die Praxis

Reihe herausgegeben von
Claus Emmelmann, Hamburg, Deutschland

Technologie- und Wissenstransfer für die photonische Industrie ist der Inhalt dieser Buchreihe. Der Herausgeber leitet das Institut für Laser- und Anlagensystemtechnik an der Technischen Universität Hamburg sowie die Fraunhofer-Einrichtung für Additive Produktionstechnologien IAPT. Die Inhalte eröffnen den Lesern in der Forschung und in Unternehmen die Möglichkeit, innovative Produkte und Prozesse zu erkennen und so ihre Wettbewerbsfähigkeit nachhaltig zu stärken. Die Kenntnisse dienen der Weiterbildung von Ingenieuren und Multiplikatoren für die Produktentwicklung sowie die Produktions- und Lasertechnik, sie beinhalten die Entwicklung lasergestützter Produktionstechnologien und der Qualitätssicherung von Laserprozessen und Anlagen sowie Anleitungen für Beratungs- und Ausbildungsdienstleistungen für die Industrie.

Weitere Bände in der Reihe http://www.springer.com/series/13397

Juan Pablo Calderón Urbina

Efficient material laser beam ablation with a picosecond laser

Springer Vieweg

Juan Pablo Calderón Urbina
Institute of Laser and System Technologies
Hamburg University of Technology
Hamburg, Germany

ISSN 2522-8447 ISSN 2522-8455 (electronic)
Light Engineering für die Praxis
ISBN 978-3-662-61885-1 ISBN 978-3-662-61886-8 (eBook)
https://doi.org/10.1007/978-3-662-61886-8

This Springer Vieweg imprint is published by the registered company Springer-Verlag GmbH, DE part of Springer Nature.
The registered company address is: Heidelberger Platz 3, 14197 Berlin, Germany

Acknowledgements

The presented work is the result of my engagement as a doctoral fellow and research assistant in the Institute of Laser and System Technologies (Instutit für Laser- und Analagensystemetechnik, iLAS) at the Hamburg University of Technology (Technische Universität Hamburg, TUHH).

This doctorate represents two very important aspects for me; the first, a scientific effort in a prestigious institution and in an admirable social, progressive and conscious country; and second, a part of my personal life. It includes professional aspects of my career and private events in my heart. It is a peak and a valley of experiences bringing joy and sweat. Sometimes because of it and others only as a witness of my own timeline. Therefore, there is a list of people deserving my respect, admiration and appreciation that have shaped the moments, decisions, experiences and heart.

Firstly, I would like to express my sincere gratitude to Prof. Dr.-Ing. Claus Emmelmann for his guidance, motivation, advice, supervision and making it possible for me a engage in high-level research and European technological collaborations. Likewise, I would like to thank Prof. Dr.-Ing. Wolfgang Hintze, from the Institute of Production Management and Technology at the TUHH, for the interest in my work, his observations and assuming the role of second supervisor of my thesis and Prof. Dr.-Ing. Dieter Krause, from the Institute of Product Development and Mechanical Engineering Design, for chairing the examination board. Furthermore, I am thankful for the support of the Mexican National Council on Science and Technology (Consejo Nacional de Ciencia y Tecnología, CONACYT) for its scholarship program at the beginning of my research.

Moreover, I would like to thank all active and former colleagues from iLAS and all students that contributed to my work and made this process a proactive and pleasant journey. Specially, I would like to thank Dr.-Ing. Dirk Herzog, for his supportive idea exchange and counseling, Mr. Marco Koslowski, Mr. Franz Terborg, Marco Haß and Ms. Fei Teng for their excellent technical assistance, Mr. Hendrik Vogel and Dr.-Ing. Eric Wycisk for their friendship, never ending help and advice, Mr. Hendrik Schonefeld for his empathetic energy, Mrs. Angela Steffen, Mrs. Martina Dorfner and Prof. Dr. Ing. Maren Petersen for their guidance and support. Similarly, with many thanks to Mr. Chiragkumar Maheshbhai Patel for his brilliant engagement as one of my students, Dr. rer. nat. Francesca Moglia and Dr. Hüseyin Çankaya from DESY (Deutsche Elektronen-Synchrotron, Hamburg) for their kindness, opening their doors and allowing me to use their experimental laboratory lasers.

Finally, from a deep place of my heart, I would like to mention my infinite gratitude to my family in Mexico, especially to my father for believing in me and supporting me from the very beginning, to my mother for giving me the values that made me come through all challenges, to my parents in-law, Antonio and Margarita, for their unconditional affection, support and generosity. To conclude, I would like to extend a very special acknowledgement to my best friend and wife *Mónica* for her support, patience, critique, love and walking our paths together; in the same way, to the *light* of my life Maximiliano and to my little *spark* of love Yunuén for giving my life intensity, coherence, an uniform wave and direction.

June 2020, Hamburg, Germany Juan Pablo Calderón Urbina

Summary

Ultra-short pulse laser processing of ultra-hard materials requires an accurate and agile experimental and analytical investigation to determine an efficient choice of parameters and settings to optimize ablation. Therefore, this work presents a quality-oriented experimental approach and an analytical approach for the modeling and validation of multi-pulse picosecond laser beam ablation on cemented tungsten carbide. This work starts with a review of literature and state-of-the-art theories of four relevant areas for this research: picosecond lasers, laser beam ablation process, cemented tungsten carbide (WC) and quality-oriented tools. Subsequently, a concept for an efficient material laser beam ablation with a picosecond laser was introduced. Furthermore, two approaches for the investigation are presented from an experimental and analytical perspective, respectively. The first approach introduced a methodology for the identification of influential parameters. It executes a quality-oriented methodology based on the SWOT analysis, cause-and-effect diagram and the variable search methodology. The conclusion of the methodology gave the interaction of pulse repetition rate f and scanner speed v_s in the form of pulse overlap and track overlap *PO/TO* as the most influential parameter in the maximization of the ablation rate. The second most influential factors resulted laser beam power P_L and burst-mode *BM*. The second approach, description of the model, executes a theoretical analysis of the picosecond laser beam ablation of cemented WC by the application of the Beer-Lambert law and multi-pulse ablation modeling. The unavailable material properties were obtained by experimental investigations, like in the cases of the incubation factor ($S = 0.75$) and the reflectivity factor ($R = 0.33$). Threshold fluence for cemented WC was determined by the application of the heat transfer theory ($F_{th} = 1960$ J m^{-2}) and input power intensity *PI* was adapted to a Gaussian beam profile. At the end of the approach, power density *Gt* visualizations of a picosecond laser pulse under the five available pulse repetition rates were modeled and validated. The findings from the adaptation of the Beer-Lambert law acted as basis for development of the multi-pulse laser ablation model for both single-pulse mode and burst-mode, respectively. Based on the definition of the number of pulses N irradiating the same area, the corresponding threshold fluence F_{th} for N, the input fluence F and incubation factor, ablation depth Z_{abl} was modeled and experimentally validated. Finally, results and conclusions of both approaches were discussed and a framework for an efficient laser beam ablation was presented. Recommendations for further actions on research and industry were introduced at the end of the work.

Table of contents

List of figures

List of tables

Abbreviations and symbols

Abbreviations

APDC	Act, plan, do and check (Deming cycle)
AOM	Acoustic-optic modulator
APM	Additive pulse mode-locking
APT	Attosecond pulse trains
CBN	Cubic boron nitride
Co	Cobalt
CPA	Chirped pulse amplification
CPM	Colliding-pulse mode-locked laser
Cr:forsterite	Chromium-Doped Forsterite Crystal
Cr:LiS(G)AF	Chromium-doped lithium strontium aluminum fluoride
Cr:YAG	Chromium doped Ytterium Aluminum Garnet
Cu	Copper
CVD-D	Chemical vapor deposited diamond
CW	Continuous wave
DESY	Deutsches Elektronen-Synchrotron
DoE	Design of experiments
EOM	Electro-optic modulator
Er	Erbium
HAZ	Heat-affected zones
HHG	High harmonic generation
IAP	Isolated attosecond pulses
IR	Infrared
KLM	Kerr lens mode-locking
LAM	Laser additive manufacturing
ln	natural logarithm
Nd	Neodymium
Nd:YAG	neodymium-doped yttrium aluminum garnet

Nd:YLF	Neodymium-doped yttrium lithium fluoride
$Nd^3:YVO_4$	Neodymium-doped Yttrium Vanadate
PCBN	polycrystalline cubic boron nitride
PCD	Polycrystalline diamond
RGA	Regenerative amplifiers
SESAM	Semiconductor saturable absorber
SLM	Spatial light modulator
SWOT	Strengths, weaknesses, opportunities and threats
TEM	Transverse electromagnetic mode
Ti	Titanium
TTM	Two-temperature model
UV	Ultraviolet
W	Tungsten
WC	Tungsten carbide
XUV	Soft-X-rays
Yb:YAG	Ytterbium-doped yttrium aluminum garnet

Latin symbols

AR	$mm^3 s^{-1}$	Ablation rate
B_1	$mm^3 s^{-1}$	First Best Qn+ preliminary experiment result
B_2	$mm^3 s^{-1}$	Second Best Qn+ preliminary experiment result
c	$m s^{-1}$	Speed of light in vacuum
C_e	$J k^{-1}$	Heat capacity of the electron system
c_p	$J k^{-1}$	Heat capacity (specific heat)
$c_{p,l}$	$J k^{-1}$	Heat capacity (liquid-state)
$c_{p,s}$	$J k^{-1}$	Heat capacity (solid-state)
CS	Mpa	Compressive strength
CTE	$10^{-6} K^{-1}$	Linear thermal expansion coefficient
d	m	Diffusion depth
D	m	Diameter (width) of ablated areas

D_s	-	Total effect due to simultaneous variation (Shainin)
d_s	-	Dispersion between repeat repetitions (Shainin)
dT	K	Change in the temperature
dt	s	Time period
d_w	m	Focal diameter
E_0	J	Pulse energy
EM	GPa	Elastic modulus
E_p	J	Input energy
ER	$\mu\Omega$ cm	Electrical resistivity
Er_m	%	Error of the model when compared to the experiment
E_{th}	J	Energy threshold
f	kHz	Pulse repetition rate
F	J cm^{-2}	Average input laser fluence
F_0	J cm^{-2}	Laser peak fluence
f_d	m	Focal distance
F_{th}	J cm^{-2}	Threshold fluence
Gt	W m^{-3}	Heat source distribution
HV	HV	Hardness (Vickers)
I	W cm^{-2}	Intensity
$I(t)$	W m^{-2}	Laser pulse intensity through time
I_0	W cm^{-2}	Threshold Intensity
k	W m^{-1} k^{-1}	Thermal conductivity
L	m	Length
LH_e	J Kg^{-1}	Latent heat of evaporation
LH_f	J Kg^{-1}	Latent heat of fusion
M^2	-	Beam quality
n	-	Mode (single/burst)
N	-	Number of applied pulses
n_L	-	Number of irradiation layers
n_T	-	Number of pulse tracks

P_I	W m^{-2}	Gaussian input power intensity
P_L	W	Laser beam power
$P_{L,0}$	W	Laser beam power emitted from the source
$P_{L,R}$	W	Laser beam power reflected from the sample
PO	%	Pulse overlap
$Qn-$	mm^3 s^{-1}	Quantitative negative setting level
$Qn+$	mm^3 s^{-1}	Quantitative positive setting level
R	-	Reflectivity
r	m	Radius
S	-	Incubation factor
$ScSt$	-	Scanning strategy
t	s	Time
T_b	K	Boiling point
T_e	K	Temperature of the electron system
t_{ee}	s	Thermalization time of the electrons
t_{ep}	s	Interaction time of the electron and phonon system
$t_{\gamma e}$	s	Absorption time by quasi-free electrons
T_m	K	Melting point
TO	%	Track overlap
t_p	s	Pulse length of the emitted radiation
T_p	K	Temperature of the phonon system
t_{pp}	s	Thermalization time of the phonon system
T_r	°C	Temperature extern (room enviroment)
TS	MPa	Tensile strength
$V-$	mm^3 s^{-1}	Negative reference value (Shainin methodology)
V_+	mm^3 s^{-1}	Positive reference value (Shainin methodology)
v_s	m s^{-1}	Scanner speed
W	m	Width
W_1	mm^3 s^{-1}	First Worst Qn- preliminary experiment result
W_2	mm^3 s^{-1}	Second Worst Qn- preliminary experiment result

y	m	Beer-Lambert law depth in a material
Z_{abl}	m	Ablation depth
z_f	m	Focus position
z_{fa}	m	Focus adjustment
z_R	m	Rayleigh range
$z_{R,\ Gaussian}$	m	Rayleigh range in a Gaussian distribution of beam
$z_{R,\ real}$	m	Rayleigh range including beam quality factor M^2

Greek symbols

α	-	Absorption coefficient
γ	°	Incidence angle
δ	m	Optical penetration depth
Δ_y	m	Hatch distance
ε_v	$W\ m^{-3}$	Power density
$\varepsilon_{v\ l\text{-}b.p.}$	$W\ m^{-3}$	Power density to reach boiling point from liquid-state
$\varepsilon_{v\ l\text{-}v}$	$W\ m^{-3}$	Power density to change state from liquid to vaporous
$\varepsilon_{v\ s\text{-}l}$	$W\ m^{-3}$	Power density to change state from solid to liquid
$\varepsilon_{v\ s\text{-}m.p.}$	$W\ m^{-3}$	Power density to reach melting point from solid-state
$\varepsilon_{v\ T}$	$W\ m^{-3}$	Total power density
θ	°	Angle of reflection
κ	-	Absorption index
λ	m	Wavelength
π	-	Pi constant
ρ	$kg\ m^{-3}$	Mass density
ρ_l	$kg\ m^{-3}$	Mass density (liquid-state)
ρ_s	$kg\ m^{-3}$	Mass density (solid-state)
T	-	Transmittance
τ_H	s	Pulse length

τ_P	s	Pulse period
ϕ^2	-	Slope of a curve
φ	°	Beam widening angle
ω_0	m	Laser beam radius

1 Introduction

The development of ultra-short pulse lasers has given laser beam ablation a wider application perspective by granting minimal thermal side-effects, industrial robustness and precision [1; 2; 3; 4]. A rising field of opportunity is the processing of hard and ultra-hard materials such as cemented tungsten carbide (WC) and polycrystalline diamond (PCD), among other purposes [5; 6; 7; 8]. The intersection of reliable ultra-short pulse laser sources, like picosecond lasers [9], and the increasing trend of hardness of ultra-hard materials [10] represents an attractive opportunity for the application of laser beam ablation [11]. This synergy, picosecond laser ablation of ultra-hard materials, must consider innovations and adapt to market and production conditions to maintain competitive results.

The actual performance of the international manufacturing arena demands the deft application of the highest competitive standards, which are based on the maximization of the quality of processes and products and the minimization of production time, delivery and costs [12]. Moreover, the continuous change of market conditions and the development of technology require flexible systems able to comply with new quality aspects and technological developments [13].

Picosecond laser beam ablation is a pyrolytic process on the material that consists of the heating, melting and evaporating of molecules of a solid by means of optical energy induced processes [14; 15; 16]. High power density is applied on the material in a very short interaction time of several picoseconds [2; 17]. Therefore, the time of energy absorption of ultra-short pulses is short enough to achieve material removal and avoid negative thermal collateral-effects, e.g., plasma formation, micro cracks, burns and undesired high roughness [17]. The challenge starts when the benefits of ultra-short pulse laser beam ablation are wanted to be transferred to other materials and objectives, like the maximization of the ablation depth, ablation rate and the minimization of the surface roughness, because this assumes the selection of the correct parameters, levels and

J. P. Calderón Urbina, *Efficient material laser beam ablation with a picosecond laser*, Light Engineering für die Praxis, https://doi.org/10.1007/978-3-662-61886-8_1

conditions for a selected material, that could be homogenous or heterogeneous, like for instance, WC, and its successful experimental data collection, as illustrated in figure 1.

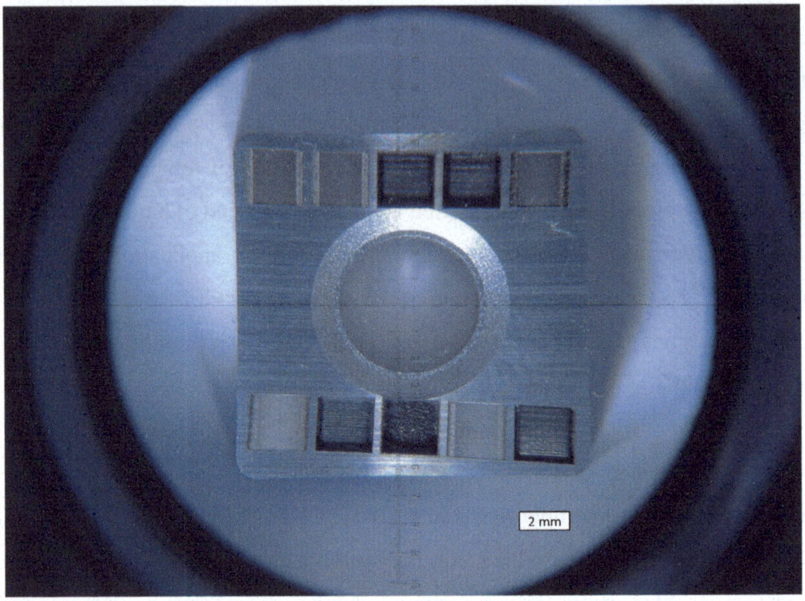

Figure 1: Picosecond laser beam ablated areas on a WC sample under investigation

The objectives of this work are, first, the scientific investigation of laser beam ablation of tungsten carbide by using a picosecond laser, and second, the creation and implementation of a systematic quality-oriented approach for the analysis of laser beam ablation based on the Deming cycle (also known as APDC cycle), depicted in figure 2, to assure quality in the process. The objective of the cycle is to achieve continuous improvement by mastering four phases. The first, *plan*, sets the basis for the operation; the second, *do*, performs the plans; the third, *check*, applies a revision of actions and plans; and the fourth and last, *act*, performs the adequate actions to comply with the objectives (plans) and to generate continuous improvement.

Based on that, this work pursues the systematic identification, description and quantification of significant influencing parameters on the picosecond laser beam ablation of WC in terms of ablation rate. The scientific application of quality-oriented tools and theoretical analyses gives an empirical-analytical strategy that avoids unnecessary research steps and saves resources efficiently (more with less) not only in the experimentation but also in the implementation.

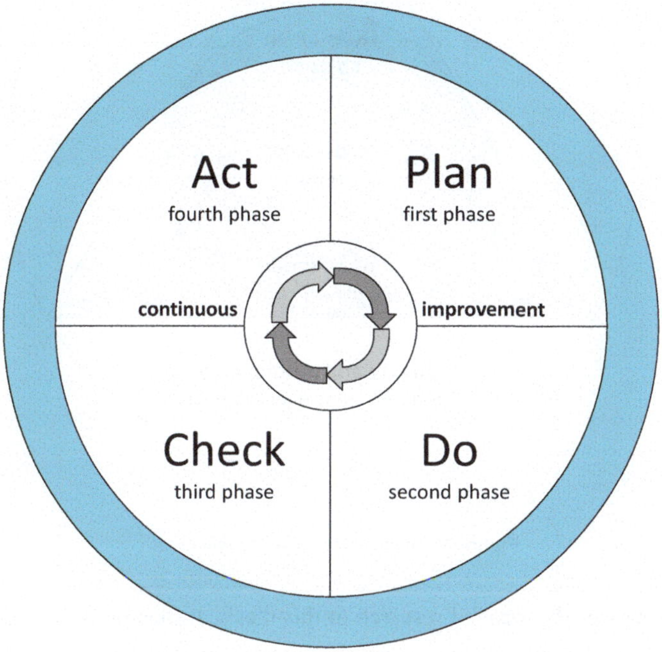

Figure 2: APDC cycle of Deming [12]

Finally, the implementation of the approach must be tested and validated under economic and qualitative production standards to be able to achieve an industrial application by assuring high quality, efficiency and process stability.

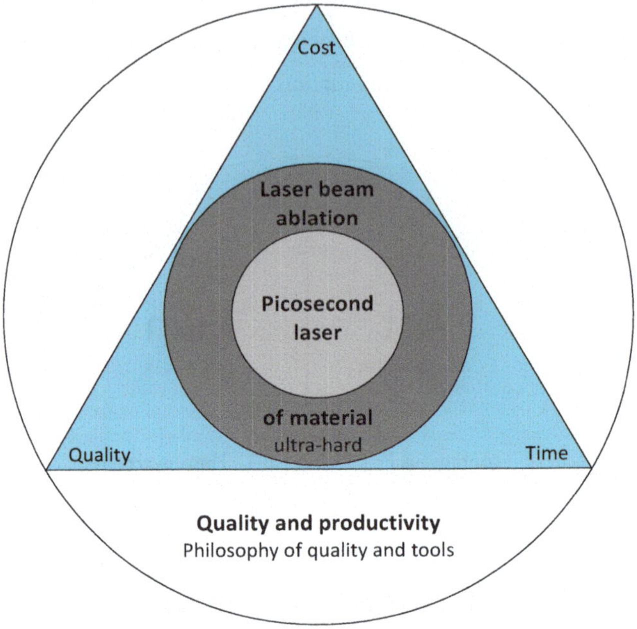

Figure 3: Intersection of the thesis areas

Figure 3 summarizes the areas of research of this thesis. It illustrates the application of a picosecond laser in the process of laser beam ablation of ultra-hard materials. The process follows the guidelines of competitiveness: better quality, lower costs and shorter times by executing a philosophy of quality and productivity with the help of quality-oriented tools.

Immersed in the above-mentioned areas and adopting the philosophy of Deming to assure quality, this work has been structured in four phases and it is described in detail in the following lines (see figure 4).

After the introducing chapter, the first phase of the cycle, *plan*, is found. It describes the theoretical basis of the work and the strategy to be executed in the following phases. Here, chapter 2 presents the state of the art and includes core aspects of the research. They describe the areas shown in figure 3. Hence, subchapter 2.1 describes ultra-short pulse lasers, their definition, types and configurations. Picosecond lasers represent a central part of this chapter.

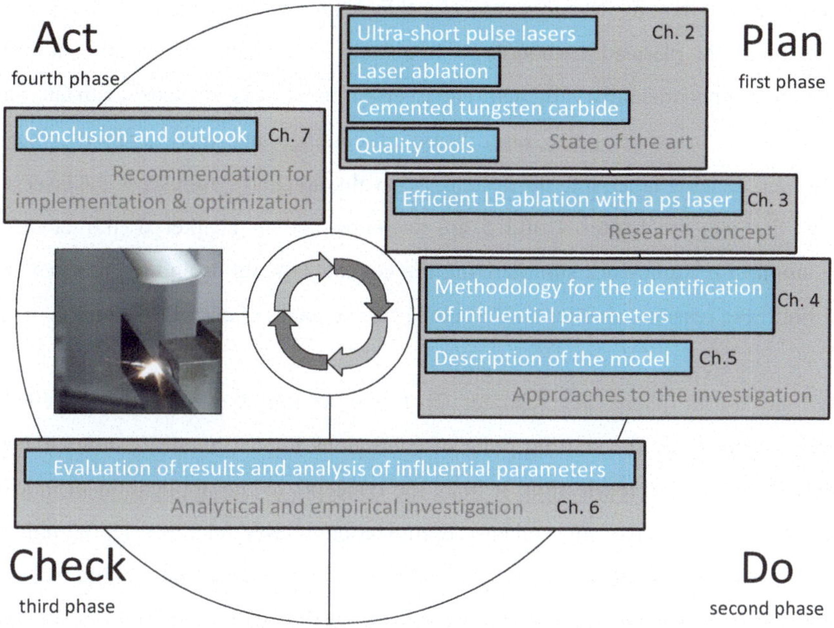

Figure 4: Structure of the thesis

Subchapter 2.2 explains the process of laser beam ablation of materials with a focus in ultra-short pulses. Subchapter 2.3 introduces the material choice for this work and thus presents a description of cemented WC, main aspects and characteristics. The quality philosophy and especially the quality-oriented tools are described in subchapter 2.4. It introduces their concepts, methods and benefits in the application in manufacturing processes. Chapters 3 and 4 introduce the main concept of the work and the plan of the experimental approach, respectively. Chapter 3 focuses on the efficient laser beam ablation using a picosecond laser. It defines the general strategy to achieve optimal results on the selected material. The definition of the methodology for the identification of significant influential parameters on the ablation process, i.e., ablation rate, is discussed in chapter 4. This chapter prepares the experimental plan for design of experiments (DoE), such as variable search methodology or Shainin´s method, executes the methodology and presents its conclusions.

Chapter 5 and 6 belong to the phase *do*; they execute, first, theoretically and then experimentally the planned methodology. In chapter 5, an analytical model is described, executed and experimentally validated. The Beer-lambert law is proposed as an alternative to understand ultra-short pulse laser beam processing, and considerations, like the incubation effect, are taken into the system. An evaluation and synthesis of the laser beam ablation findings in chapter 4 and 5 are accomplished in chapter 6, it includes the presentation of a framework summarizing the concepts, methods and approaches of the work. For these reasons, chapter 6 also belongs to the phase *check* of this work.

Lastly, the fourth phase *act* (chapter 7) refers in this thesis to the conclusions, recommendations and the starting point of continuous improvement. It summarizes the findings for industrial application; it links the concepts to the application in industrial aspects in terms of quality, time and cost (optimization of laser ablation). The combination of optimal parameters and levels, the flexibility to adapt to changing conditions and the integration to actual laser machines represent an attractive methodology for the operation of laser beam ablation technologies. Thus, this last chapter summarizes the work and presents general conclusions of the thesis. Besides, fields of improvement and lines of investigation for future works in the topics of ultra-short pulse laser beam ablation are presented (outlook).

2 State of the art

2.1 Ultra-short pulse lasers

The second chapter of the thesis is placed in the first phase *plan* of the systematic quality-oriented approach of the thesis. Figure 4 shows the detailed position of the subchapters of the state of the art in the structure of the thesis.

In general terms, laser beam micromachining is the result of a laser beam acting on a selected material for a defined objective. In this case, the contact of the laser beam on the material contains photons which energy is absorbed by the electrons and causes the increment of the temperature. For this reason, the lattice gets heated by electron-phonon collisions and will continue the process of heating up in dependence of the electron-phonon coupling and the length of the laser beam pulse [18]. According to the length of the pulse, there is a different laser beam material interaction with long and short pulses. As a matter of fact, thermal diffusion plays a greater influence when interacting with long pulse durations; it heats a greater part of the lattice reaching its melting temperature and then its vaporization temperature when enough power is available. This causes the melting of the material and a small amount of vaporization [2; 18]. In contrast, short laser beam pulses deliver in a shorter interaction time an intensity that causes high electron temperatures. Thus, the electron-phonon coupling is executed after the pulse and it contributes with high lattice temperatures that overpassed the ones of longer pulses using the same energy. The concentration of the energy in the interaction area presents a short melting phase going faster to the vaporization temperature causing much of the material in the zone to vaporize [18]. Ultra-short pulses accelerate the laser beam-material interaction by reducing thermal properties of longer pulses. For example, the thermal conduction and thermal diffusion are minimized causing more emphasis in the photochemical processes and like this taking advantage of the minimal thermal side-

J. P. Calderón Urbina, *Efficient material laser beam ablation with a picosecond laser*, Light Engineering für die Praxis, https://doi.org/10.1007/978-3-662-61886-8_2

effects in the ablation process [2]. The next part 2.1 explains relevant consideration of ultra-short pulses.

2.1.1 Definition – When a pulse is considered short and ultra-short

The development of laser pulses has achieved a shortening of pulse length during the last decades which has represented (and represents) new possibilities of material processing. This trend of minimizing the pulse length has helped to process materials that had been difficult and costly to process with longer pulses and with traditional technologies. An example of this is the processing of hard materials [5; 7; 9; 19]. For this reason, it is important to understand the concept of ultra-short pulse and how they are classified like this when compared to other pulses.

The classification of short and ultra-short pulses will be explained in this work from the perspective of the material and based on Dirscherl, Meijer et al [2; 14; 16; 18]. During laser-material interaction a beam of photons is absorbed in the outer layer or skin layer of the material by its free electrons in about a time of 1 fs, 1 x10^{-15} s, and according to optical characteristics of material and laser. Then, the energy is kept in the electrons during their relaxation time for about 1 ps, 1 x10^{-12} s, until this time is over; it is then transformed into heat [2; 14; 16]. There are materials that lack of free electrons and would not fit for material interaction with ultra-short pulse lasers from an initial point of view. Nevertheless, the energy of ultra-short pulses is successfully received on the material by multi-photon absorption, electron impact absorption or interband transitions [20]. This mechanism of ultra-short pulses makes possible the processing of this kind of materials, such as, ultra-hard and transparent materials.

Due to the optical properties of the beam and the material, the intensity of the laser beam I_0 is reduced by the depth of the material and expressed in equation 1:

$$I_x = I_0 e^{-\alpha x} \quad (1)$$

Here, α is the optical absorptivity coefficient of the material and x the depth through the material. The optical penetration depth δ describes a distance in which almost all the laser beam energy is absorbed into the skin depth and it is conditioned by α and calculated by:

$$\delta = \frac{2}{\alpha} \tag{2}$$

The relaxation time of electrons in metals after exposed to a laser beam is of 1 ps. During this time, the optical penetration depth is of approximately 10 nm that are being heated by the laser beam energy. Hence, this caused heat is transported or, specifically, diffused from the outer layer to the interior of the material and is defined by the diffusion depth d:

$$d = \sqrt{4at} \tag{3}$$

Where, a is the thermal diffusivity (also known as coefficient of diffusion) in a diffusion time t into the material. The authors [16] exemplify this notion with steel. The diffusion depth for this material corresponds to 1 nm when a pulse of 10 fs impacts its surface. Likewise, $d = 10$ nm, when a 1 ps pulse impacts the material. In other words, when a laser beam interacts with the selected material, it lasts 1 ps to transform the laser beam energy into heat and penetrates 10 nm into the material in its skin layer, and consequently it presents a diffusion depth d of 10 nm for a pulse of 1 ps.

From all these considerations, it can be said that a pulse is ultra-short when the thermal diffusion depth d is in the same magnitude or less than the optical penetration depth δ during a pulse interaction, i.e., $d \leq \delta$. The thermal diffusion depth is dependent of the material properties and the optical penetration depth depends on the optical properties of the materials and the optical properties of the laser beam like its wavelength λ. Figure 5 illustrates ultra-short laser-material interaction and its related considerations.

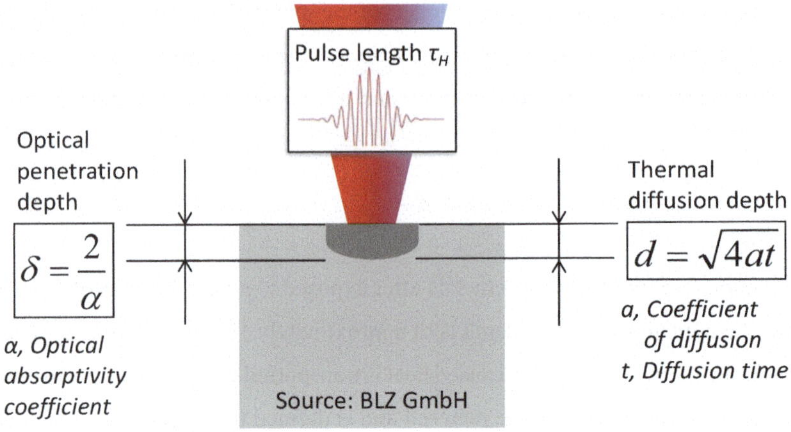

Figure 5: Ultra-short laser beam-material interaction [2]

The material properties and its optical characteristics, such as thermal conductivity, reflectivity, and absorption, define the scope of ultra-short pulses; therefore, the limit between short and ultra-short pulses is of approximately 1 ps for metals, 10 ps for ceramics and 1ns for certain polymers due to their low thermal conductivity. For this work, a picosecond laser is selected. It delivers a pulse length of 10 ps and it is considered as an ultra-short pulse laser. In the following subchapters, it is explained into detail.

2.1.2 Working principle and configuration of ultra-short pulse lasers

The simplest method to obtain pulsed laser beams, as mentioned before, corresponds to the input of pulsed pumping energy by flashing lamps. This configuration reaches the emission of pulsed laser beams in the range of microseconds (μs). The reduction of the pulse length has been achieved thanks to the application of other technologies such as Q-switching and mode locking that simultaneously have been complemented with such components as Kerr-lens mode locking and CPA-technology to improve their performance.

Figure 6 presents the relationship and development between pulse length and its corresponding technology. It can be also observed that the rise of ps pulses has its origin from the development of mode-locking. Therefore, the understanding of mode-locking for ultra-short pulse generation requires a greater focus and hence this part explains into a detailed perspective this relevant configuration.

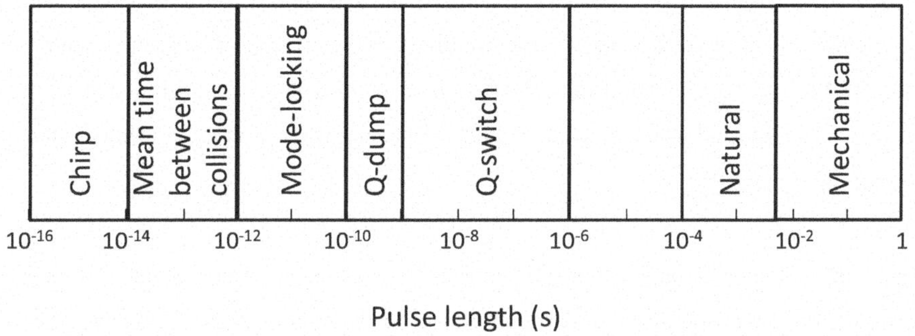

Figure 6: Time scales of pulsed laser output [21]

Mode-locking

A laser beam consists of two relevant modes that help to describe its beam behavior and quality and to characterize its output. One of them is the transverse beam mode and the other is the axial or longitudinal beam mode. The transverse mode describes the distribution of the laser intensity within the cross section of the beam, i.e., in a plane perpendicular to the propagation direction of the laser beam; this mode supports the characterization of the beam quality by, e.g., the TEM (transverse electromagnetic mode) classification [18; 22]. The second mode, axial, corresponds to the waves along the longitudinal axis created, amplified and affected under certain conditions (cavity design, medium, losses, etc.) in the cavity after the oscillation of the electromagnetic field [2; 18]. These waves or longitudinal modes in the cavity oscillate independently among them; thus, they differ in their field distribution along the x axis, amplitude and phase due to the fluctuations of the resonator length and the nonlinear interaction between the modes [18; 23].

The behavior and control of the oscillations of the longitudinal modes represent the key focus of the mode-locking technology.

Based on the fact that a laser cavity consists of several oscillating modes behaving differently and independently from each other [18], it is produced a random distribution of mode phases and an apparent constant average emission of a superimposed continuous laser beam (see figure 7) [2]. However, if the oscillating modes are forced or locked to swing under a fixed phase by means of a modulator and start to oscillate at the same time [2; 18; 21; 23; 24; 25], their mode amplitudes reach a constructive interference at a certain point in the cavity (and a destructive in almost the rest of it) producing a high intensity and short pulse [2; 25].

Figure 8 illustrates schematically the lock of three modes. The modulation of the cavity losses with the roundtrip time, and the attention to dispersion interactions are relevant topics for the locking of the modes [24].

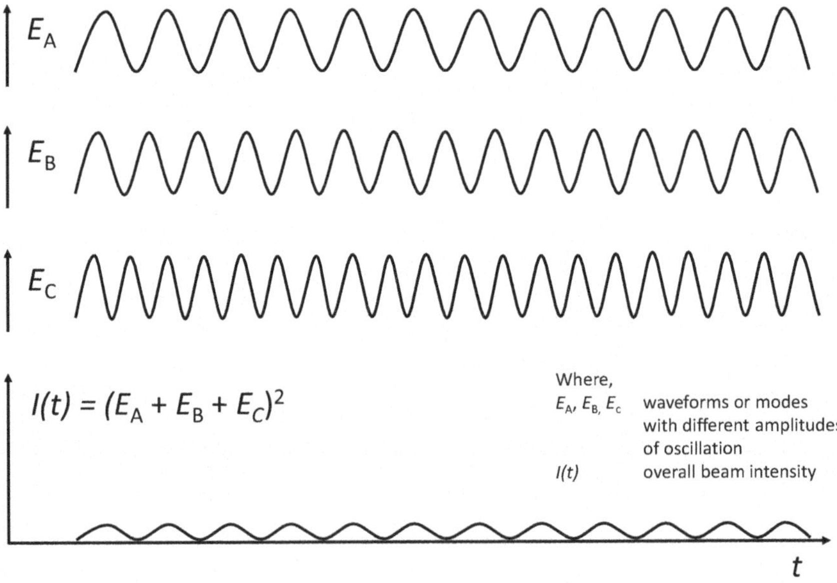

Figure 7: Superposition of three equally spaced waveforms with different phases [18]

In this technique, **mode-locking**, the mentioned short pulse oscillates in the cavity and produces an output short pulse every time it passes through the output main mirror, contrasting like this with the longer (cavity) round-trip time [25].

The length of the pulse is based on the fact that the oscillating mode-locked bandwidth maintains an inversely proportional relationship to the length or duration of the output mode-locking pulse [2; 18; 25].

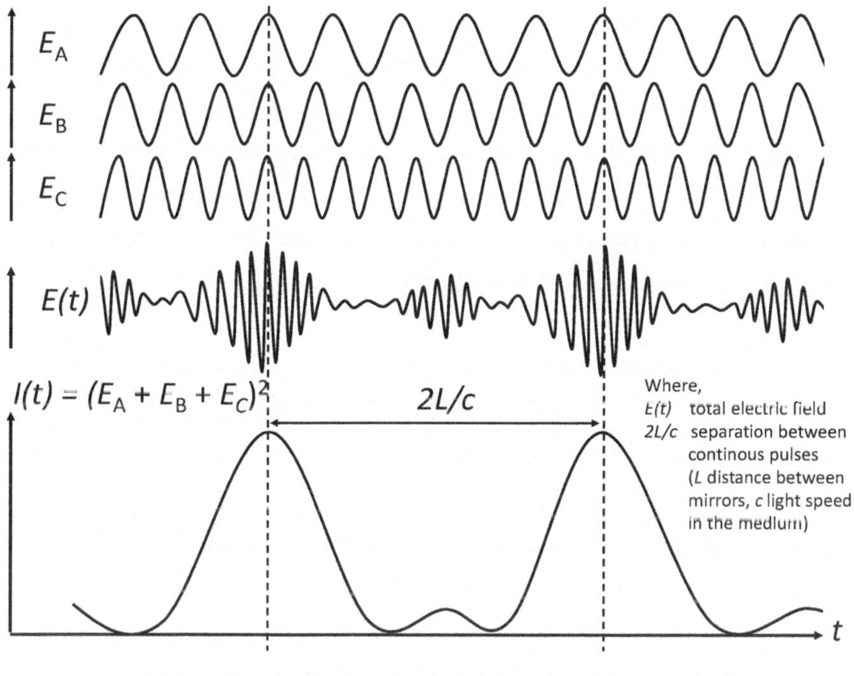

Figure 8: Superposition of three equally spaced waveforms of the same phase [2; 18]

Thus, the emission of shorter pulses requires several oscillating modes, and consequently, a larger bandwidth [2].

Mode-locking technology makes feasible the emission of pulses of short duration, such as nanosecond, picosecond and femtosecond [18; 24], only if the spectral bandwidth of the active medium is capable of having the required axial modes [24].

The typical achieved pulse length with mode-locking in lasers having gas as an active medium is of nanosecond range [18; 23]. The length for lasers using solid-state active mediums goes from picosecond to femtosecond durations (e.g., by the use of dye lasers) [18; 23]. The intensity of the laser medium excitation determines the number of the axial modes in the output of a beam and the number of wavelengths [21]. A beam of a single wavelength is achieved by an excitation of the medium slightly over its threshold value; dissimilarly, beams of various wavelengths can be obtained by a significant excitation of the medium over its threshold value [21].

Mode-locking technology offers pulses of a high peak power of the order of gigawatts [18]; however, this could not be achieved if a fundamental transverse cavity mode TEM_{00} is not available because higher order transverse modes keeping high energy levels have different resonance frequencies that disable a periodic recurrence of constructive interaction of all mode amplitudes; based on this, a TEM_{00} is a relevant condition for a reliable and stable mode-locked operation [25].

Mode-locking has been present since the beginning of laser itself. Solid-state lasers were the active medium to achieve groups of pulses or bursts in microsecond and in a frequency of Kilohertz in the 1960´s [25]. From the 1960´s the development of mode-locking has obtained several achievements in performance, high power, reliability and pulse length, from microsecond in the 1960´s and picosecond and femtosecond in the recent years. Some of the relevant historical events of mode-locking are Q-switched mode-locking, passive mode-locking with dye lasers, CPM (colliding pulse mode-locking), KLM (Kerr lens mode-locking) and CPA (chirped-pulse amplification) among others.

Table 1 summarizes the highlights of the development of mode-locking from 1960 to 1992. Mode-locking types, such as active and passive, and their corresponding actual developments are explained in the following lines of this subchapter.

Table 1: Historical development of mode-locking [25]

Year	Historical development	Remarks
1960´s	First demonstration of mode-locking	Use of solid-state lasers for bursts of pulses in μs and kHz. No CW mode-locked operation
1960´s	Q-switched mode-locking	Caused by the saturable absorber for mode-locking. High pulse energy and peak power but instable
1974	Demonstration of sub-picosecond passive mode-locking dye lasers	
1981	Demonstration of sub-100-fs colliding pulse mode-locked dye lasers (CPM)	Pulse of 27 fs and an average power of approximately 20 mW
1980´s	Development of high power (diode) solid-state lasers Shortening of pulse length using solid-state lasers	Diode laser pumping increased efficiency, size and lifetime active mode-locking diode pumped Nd:YAG and Nd:YLF emitted pulses of 7 to 12 ps
1991	Discovery of Kerr lens mode locking (KLM)	Ti:sapphire lasers with pulse lengths less than 6 fs and without any extra pulse compression
	Appearance of chirped-pulse amplification (CPA) for Ti:sapphire systems	Supporting phase lock of the carrier-envelope-offset (CEO)
1992	Reliable design of first intracavity saturable absorber	Correction of Q-switching instabilities in solid-states lasers

There are two main classifications of mechanisms for the realization of mode-locking [18; 23; 24; 25; 26]. The first classification includes the mechanisms using active modulators, such as electro-optic and acousto-optic, and it is known as active mode-locking. The second classification, passive mode-locking, incorporates the use of saturable absorbers for the process of mode-locking. The combination of active and passive mechanism is considered by some authors as hybrid mode-locking [23; 25].

In the **active mode-locking** a modulator emits a signal to control the losses in the cavity and to maintain a desired equal relationship of the frequency and the intermodal separation, i.e., an exact synchronization of the loss modulation and the round-trips in the cavity [18; 25]. Specifically, the frequency of the axial modes' separation $c/2L$ is set by the modulator (where, L is the length of the optical cavity) and then the cycle of the frequency modulation [23]. The cycle of the frequency modulation has to correspond to round-trip time of the radiation in the cavity in order to obtain high losses of the radiation in the entire resonator and simultaneously only allow oscillation with minimum losses through the modulator when this also presents minimum losses [23]. The oscillation of a synchronized pulse in the cavity is normally not affected by the performance of the modulator because the pulse reaches the times of minimum loss of the modulator; however, the pulse tails are affected by receiving a higher loss that the pulse center, and consequently, a shortening of the pulses and the possibility of ultra-short pulses [25]. The modulator in the cavity is controlled electrically and can be either acousto-optic or electro-optic. Figure 9 illustrates schematically an active mode-locking configuration.

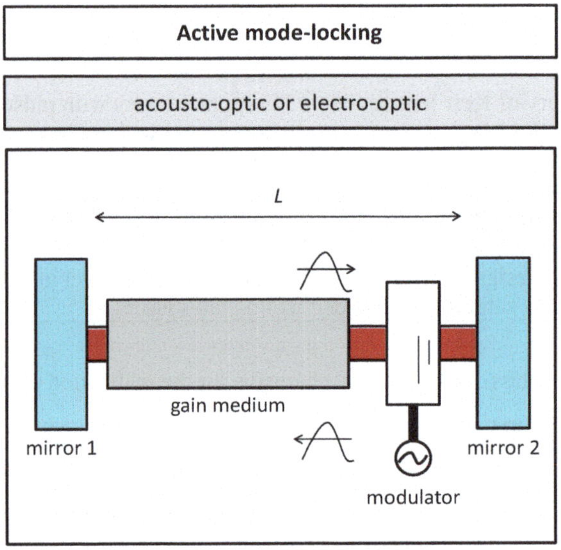

Figure 9: The schematic of the acousto-optic or electro-optic mode-locking regime [23]

The **acousto-optic modulator** (AOM) emits an acoustic wave (frequency of tenths or hundredths of MHz) in an acousto-optic medium by means of a piezoelectric transducer [25], as in Q-switching [21; 23]. The radiation is refracted by the interaction of the acoustic signal in the medium in form of a periodic index pattern and hence the refracted laser light is dispersed out of the cavity and refraction loss oscillates (single pulse propagation) [18; 25]. Another active mode-locking mechanism is the electro-optic. **Electro-optic mode-**

locking modulates the frequency by means of an electro-optic device in the cavity fed by an electric signal and causing sinusoidal fluctuating frequency shift in the radiation going through it [23]. The successful synchronization of the modulation frequency and the round-trip in the cavity causes that some radiation in the cavity presents up-shifts and down-shifts in frequency; consequently and after several cycles, the up-shifted and down-shifted radiation is eliminated of the laser and the part of the light passing through the modulator when the frequency shift equals to zero, is the one that remains uninfluenced and generating a tapered pulse of light [23]. The last active mode-locking mechanism described in this work, is the **synchronous mode-locking**. This active method, also known as synchronous pumping, uses the pumping source as a self-operated switch that turns on and off to produce laser pulses and thus modulate the oscillations [23]. Synchronous mode-locking usually requires as a pumping source another mode-locked laser and a reliable and precise matching of the cavity length of both lasers (pumping laser and operating laser) [23].

Active mode-locking systems require a precise synchronization of loss modulation and cavity round-trips; they rely on active modulators and can produce pulses in the order of picoseconds [25; 26]. On the other hand and as explained in the following lines, passive mode-locking systems are simpler, do not rely on core electric components (due to the use of saturable absorbers) and can generate pulses shorter that those from active mode-locking [25].

The use of saturable absorbers is a fundamental characteristic of **passive mode-locking**. Saturable absorbers have an increment of absorption to a low incident beam intensity and a decrement of it to a high incident beam intensity [18; 21]. The position in the cavity of the saturable absorber is similar to the configuration of Q-switching systems; it is placed next to the output main mirror [26].

In passive mode-locking, the mode-locked pulse generation starts with the emission of a not mode-locked laser beam incident to the saturable absorber. At the absorber, the incident low intensity pulses (due to intensity fluctuations of not mode-locked beams) are eliminated and the high intensity pulses bleach the absorber and increase the transmission through it (as schematically illustrated in figure 10).

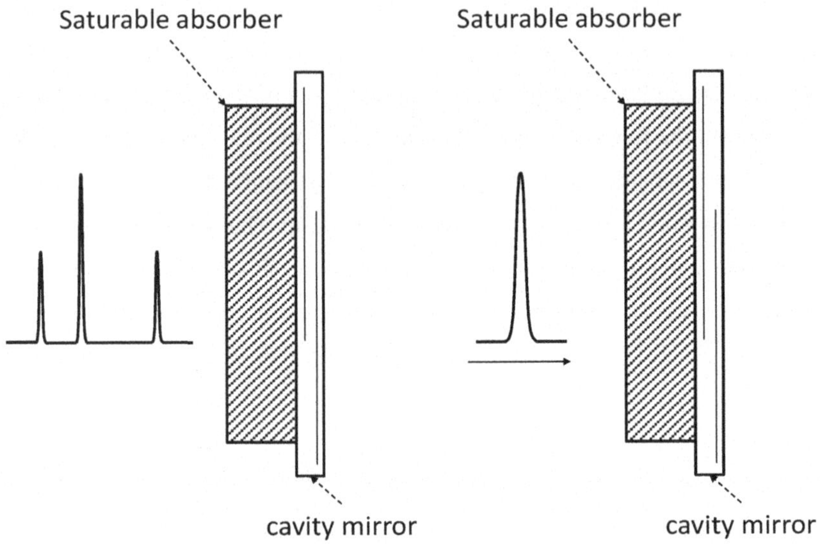

Figure 10: Passive mode-locking using a saturable absorber [27]

Consequently, the first high intensity pulses oscillate with the round-trip time in the cavity bleaching the absorber and continuing their amplification and growth thanks to the low loss [18; 25; 26]. In order to obtain an effective process, it is necessary that the recovery time of the saturable absorber is shorter than pulse length [18; 25]. The laser radiation that continues oscillating in the cavity repeats the selection process of absorption of low intensity pulses and amplification of high intensity pulses until the mode-locking of the laser is reached [23]. The output consists of a train of pulses of very short length influenced by mode-locking and Q-switching behaviors [18].

There are a high number of passive mode-locking mechanisms that are relevant for the generation of ultra-short pulses, e.g., colliding pulse mode-locking, semiconductor saturable absorber mirrors, Kerr lens mode-locking, additive pulse mode-locking, resonant passive mode-locking and nonlinear mirror mode-locking, among others [2; 23; 25]; the relevant mechanisms considered in this work are mode-locking by semiconductor absorbers, Kerr lens mode-locking, additive pulse mode-locking and nonlinear mirror mode-locking. The first of the listed mechanisms is the one using semiconductor absorbers. Semiconductors are capable to absorb and transmit light under certain energy conditions; for this reason, they are integrated into a mirror construction that complements its characteristics by increasing reflectivity when an incident optical intensity is also increased [25]. Thus, **semiconductor saturable absorber mirrors** or SESAMs offer a robust and simple configuration based on quantum-well structures [26], good mode discrimination, and are adequate for solid-state passive mode-locking thanks to a large absorber cross section that contributes to small saturation fluence [25; 28]. Using certain configurations, SESAMs can produce a pulse length in the order of 10 ps and are able to operate at pulse repetition rates up to MHz due to the dependence to the pumping source frequency [28]. Other configurations based on diverse laser mediums and different saturable absorber parameters can produce pulses of pulse length from nanosecond to few femtosecond [29]. For this reason, SESAMs can be operationally optimized in the aspects of laser wavelength, pulse length and power [25]. The second passive mode-locking mechanism explained in this work is **Kerr lens mode-locking** (KLM). The Kerr effect describes the relationship and dependence of the refractive index of a medium on the light intensity, e.g., the refractive index increases with the increment of the intensity [25; 26]. Furthermore, this effect contributes to the generation of short pulses due to its response time on the order of femtoseconds [26]. Based on those facts, Kerr lens mode-locking executes a self-focusing of a high intensity beam inside the gain medium to the laser beam center (transverse mode profile), i.e., it uses its gain medium as a Kerr medium to profit from the focusing Kerr effect [25; 26]. KLM includes an aperture in the cavity to neglect the oscillation of low energies and allow for the concentration of the laser energy to high intensity pulses [2], see figure 11. The aperture in the cavity can be either a hard aperture or a soft aperture; the first corresponds to the use of a pinhole in a determined position in the cavity to allow for a certain beam intensity, and the second, corresponds to the application of a pump volume with a smaller beam radius than the one from the gain medium to increase the overlap between both beams, and consequently, to obtain higher laser gain and shorter laser pulse length [2; 25; 26]. The combination of KLM and SESAM has made possible the generation of laser pulse lengths of approximately 6 fs (Ti:sapphire laser) thanks to pulse shaping and self-starting capabilities of KLM and SESAM, respectively [25; 26].

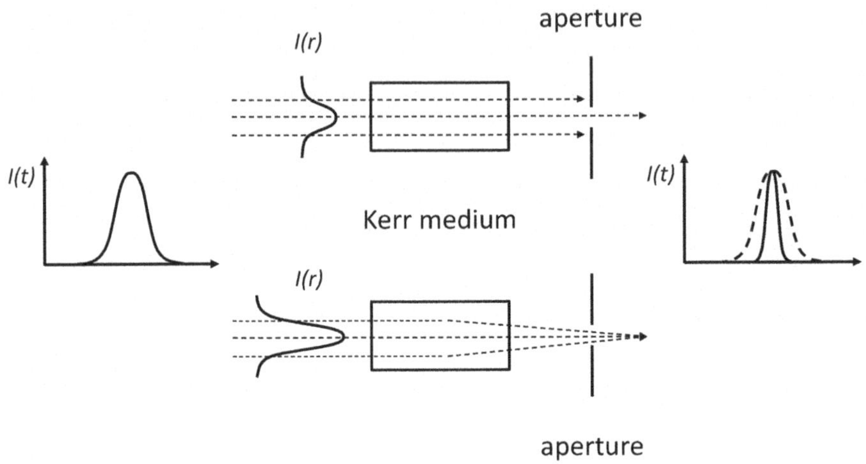

Figure 11: Kerr lens mode-locking [26]

The third passive mode-locking mechanism is called **additive pulse mode-locking** (APM) and it consists fundamentally of two pulses displaced in phases that interfere to compensate their centers and attenuate their tails [23]. In detail, APM uses the Kerr effect to produce a phase shift to the pulse center from a single-mode glass fiber in an adjacent cavity (connected to the main cavity); consequently, when the pulses of both cavities reach the coupling mirror, interference is created, and thus, the pulse center is intensified to the main cavity and their tails eliminated [25]. Even when APM requires a precise synchronization of the pulses, it has been shown to be effective in the generation of picosecond pulses (e.g., shortest pulses of 1.7 ps, Nd^{3+}:YAG) [25]. Finally, the last passive mode-locking mechanism is based on nonlinear mirrors and for this reason called **nonlinear mirror mode-locking**. These kinds of mirror benefit from nonlinearities, which behave effectively as saturable absorbers. A nonlinear mirror consists of a frequency doubling medium (crystal) and a dichroic mirror (reflection of different colors based on the direction of transmitted light) [25]. As detailed by Paschotta and Keller in [25], the incident radiation is transformed to the second harmonic, for which the nonlinear mirror reflects the most, and then, transformed back to the original beam if a precise phase shift for fundamental and second-harmonic radiation is executed; simultaneously, all fundamental radiation that was not able to be transformed is eliminated at the nonlinear mirror, i.e., the system presents high reflectivity at high intensities. The pulse length of this kind of system is determined by the interaction between fundamental and second-harmonic radiation [25].

Supporting technologies

The developments of ultra-short pulse lasers based on mode-locking started to offer a pulse repetition rate in the order of MHz; this fact combined with the available laser powers resulted in pulse energies of about few nJ. The applications, such as material processing, demanded higher pulse energies to achieve successfully their objectives. For this reason, the development of an amplification technology was necessary and the **chirped pulse amplification** (CPA) appeared in 1985 [2; 30; 31]. The capability of CPA allows increasing the energy from a system by approximately six or seven times the original energy (mJ) [2; 25]. The CPA configuration comprises a regenerative amplifier as an important part of the system that operates as an independent cavity configuration [2]. **Regenerative amplifiers** (RGA) are composed mainly by an optical cavity (figure 12), in which the pulse to be amplified is launched from a mode-locked oscillator that could also appear as an ultra-short pulse oscillator working as a **seed laser** (laser which output is injected) [2].

Once inside, the pulse exits the cavity through the polarizer after reflection from the main end mirror when the Pockels cells are deactivated [28]. If the Pockels cells are on, i.e., a quarter-wave voltage has been fed to it, the pulse is orthogonally polarized and travels back to the cavity though the polarizer without the possibility of going out. Therefore, the pulse oscillates several times through the gain medium and becomes amplified. Finally, a second quarter-wave voltage is fed to the Pockels cell and the amplified pulse is able to travel out of the cavity. Devices as a Faraday rotator ensure the direction of the amplified pulse in the system [28]. Some of the possible outputs of solid-state lasers using CPA technology reach from 1 to 10 KHz with pulse energies in the order of mJ to 100 kHz with pulse energies in the order of μJ [2].

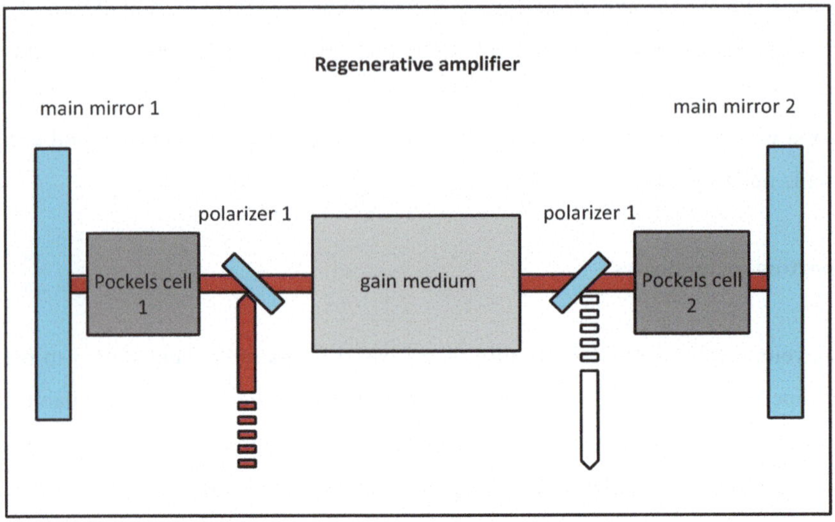

Figure 12: Schematic setup of a regenerative amplifier [2]

The possibility of ultra-short pulses, i.e., femtosecond, being injected in the cavity and passing through the amplifying crystal increases the risk that high-power density damages the components of the system. For this reason, those ultra-short pulses leaving the oscillator are extended in a grating formation (anti-parallel grating pair) to approximately thousand times of their pulse length by means of a spectral diversification [2; 26]. This temporal widening of the pulses (chirp on the pulse) decreases significantly the pulse peak power and is executed in a configuration named **stretcher**, depicted in figure 13.

After the amplification of the extended pulse and avoiding the risk of high-power density, the use of a **pulse compressor** is required. The compressor consists of an arrangement of an optical grating that compensates the effect of the stretcher and compresses the pulse to an almost similar original pulse length [2; 26]. Figure 14 shows the schematic concept of CPA technology, which includes stretching, amplification and compression.

The manipulation of the optical components of the pulse (spectral diversification) contributes to the optimization of the temporal sequence of the applications, e.g., material processing [28], by converting a laser beam pulse into a group of pulses with a constant phase relation able to interact conveniently with the workpiece to be processed or treated [28].

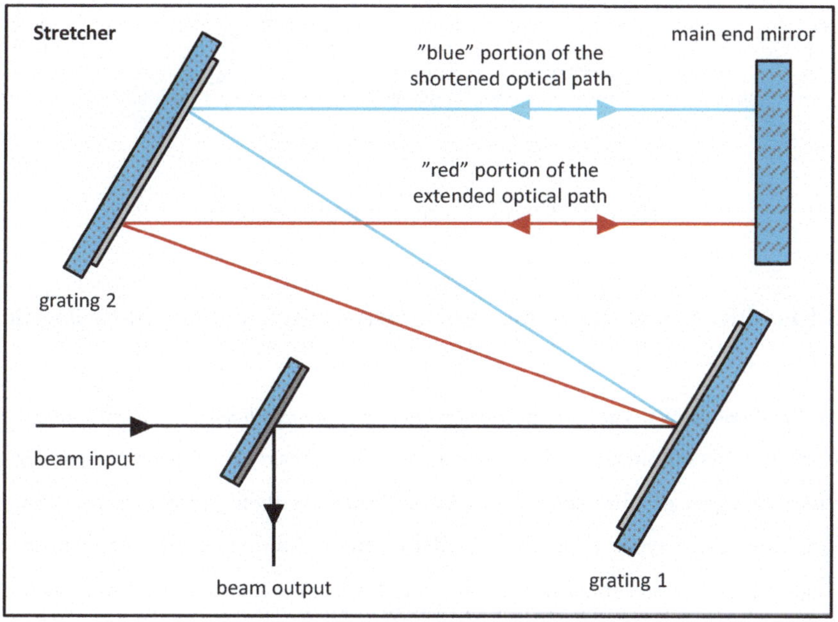

Figure 13: Schematic setup of a stretcher [2]

Gratings, like the ones used in CPA technology, allow to separate the frequency components of a pulse before entering a **spatial light modulator** or SLM, a computer-based voltage-controlled array of liquid crystalline pixels, that causes a phase differentiation in the components and with the assistance of a computer, monitoring system and control algorithms achieves specific application-oriented results [28].

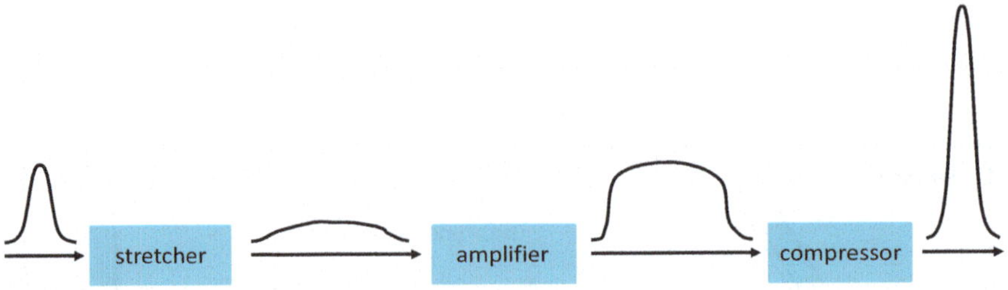

Figure 14: Schematic setup of CPA technology [2]

2.1.3 Classification of ultra-short pulse lasers based on their pulse length

One of the most relevant aspects in the selection of a pulsed laser for material processing is the length of its output pulse. For this reason, it is necessary for this work to understand the different types of ultra-short pulse lasers based on their pulse length. Thus, the description of three types of lasers considering the definition of ultra-short lasers (cp. Subchapter 2.1.1) is presented in the following lines. These are nanosecond, picosecond and femtosecond lasers; the second of them will be widely described in subchapter 2.1.4 due to its importance to this work.

Nanosecond

As known in the photonic environment, a billionth part of a second or 10^{-9} s is the unit of the pulses that are produced in a nanosecond laser. The output length of these lasers starts from 1 ns to several hundreds, e.g., 30 ns, 300, ns, 700 ns, etc.

The demonstration of Q-switching in 1961 and 1962 in a ruby laser using Kerr cell shutters marked the origin of nanosecond pulses of a length of about 10 ns [32; 33; 34]. Consequently, these kind of lasers continued their development in the late 1960´s demonstrating rise times and constant pulse widths in the order of ns and initiating the development of short pulses in industrial applications [21].

Nanosecond lasers are normally configured using the Q-switching principle, which makes possible a laser gain by avoiding oscillations in the cavity and at a certain point allowing them to generate a large intensity ns pulse [28]. Q-switching has been used not only to obtain nanosecond pulses but also to be a pioneer technology for picosecond pulses generation [35]. Other technology used to generate nanosecond pulses is gain switching; it consists in the modulation of pump power to obtain short optical pulses [36]. In detail, the gain medium is pumped to achieve population inversion thanks to the density of excited ions (through resonator round-trips); then, the optical power in the resonator reduces its intensity and causes the spontaneous emission to be amplified. This process creates a high energy peak just preceded by the off of the pumping source to avoid subsequent low-intensity pulses [36; 37]. The output pulse length is a result of cavity design, pump pulse length, repetition rate and characteristics of the gain medium [36; 37]. Nanosecond fiber lasers [37] and solid-states lasers are known to use gain-switching in current applications; however, the second ones prefer Q-switching due to the fact that they generate higher energy outputs [36].

Nanosecond lasers represent a reliable and strong alternative for applications out of experimental environments. For example, industrial-related applications like laser beam marking and engraving, laser beam ablation, laser beam cutting and laser beam drilling, among other micromachining applications, are well-established, cost-effective and applicable to a broad selection of materials [38; 39; 40; 41]. They offer a reliable processing in tasks requiring a fast and large removal of material thanks to their pyrolytic processes: heating, melting and removing the processed material [3; 42; 43]. Recent adaptation of the pulse shape and an understanding of temporal energy input in a nanosecond laser system might increase its processing efficiency, i.e., ablation, and contribute to its further application [44]. Under some applications and qualitative criteria, the application of nanosecond laser beam pulses might produce a large heat-affected zone (HAZ) that does not satisfy the goals, i.e., clean edges, minimal recast, damage-free material neighbor areas, lack of delamination, etc., and thus, a shorter pulse length might be required [42; 45].

Picosecond

A matter of great importance for this work is the study of these kind of lasers. Pulse lengths in the range of picoseconds (ps) correspond to one or more trillionth parts of a second, i.e., 1 ps equals to 10^{-12} s. Subchapter 2.1.4 addresses deeply the corresponding topics of picosecond pulse lasers.

Femtosecond

The achievement of passive mode-locking generating picosecond pulses, a background explained in the following subchapter 2.1.4, prepared the field of the generation and development of femtosecond pulses, pulses having a length of a quadrillionth part or parts of a second (10^{-15}). In 1972 passive mode-locking of a continuous wave (cw) dye laser was achieved and it described the conditions for pulse formation of this kind of cw lasers [46; 47; 48]. The conditions consisted in the relationship between saturable gain, saturable absorption and their balance determining the pulse length, and thus, opening the possibility of minimizing the pulse length in shorter degrees than the characteristic response times of both absorbing and emitting dyes [31; 34; 49]. The first appearance of femtosecond pulses occurred in 1974 and consisted of pulses shorter that 1 ps [34]. They were generated using passive mode-locking of a dye laser, where a tunable broad-gain laser medium and a saturable absorber were placed close to the center of the resonator in order to target a favorable relationship between saturable gain and saturable absorption [49; 50]. Further developments, like colliding-pulse mode-locked laser CPM in 1981 (subchapter 2.1.2), set the pulse length below 100 fs [31; 51]. The 1980`s was the decade of rise of femtosecond lasers and this event represented a prelude to the incorporation of ultra-short pulse lasers to the automotive and aerospace industries thanks to the parallel development of integrated laser systems [21]. The contemporary consolidation of femtosecond lasers is related to the demonstration of Peter Moulton in 1982 at the Massachusetts Institute of Technology. He achieved a continuous laser output emitted from a titanium-doped sapphire, analyzed and described in 1986 ($Ti:Al_2O_3$), and brought it to the commercial market in 1988 [21; 52; 53]. Ti:sapphire laser became a reliable medium for the generation of outputs of only few femtosecond pulses. Another relevant development was the demonstration of CPA technology in 1985; the "compression of amplified chirped optical pulses" [30] considerably contributed to the handling of pulse length and pulse energy in the femtosecond regime. In 1987 pulses of 6 fs were achieved experimentally thanks to a configuration of prims and gratings that compensated the optical phases [49; 54]. 1988 brought APM (subchapter mode-locking), a technique using an external cavity to achieve self-phase modulation, phase shift of the peak and the wings of the pulse, and reinjection of partial pulses to the cavity that constructively interact with following cavity pulses and destructively with their wings to achieved shorter pulses [31; 34; 55]. APM challenged the development and improvement of saturable absorbers.

In the 1990`s ultra-short pulses, such as ps and fs, were deeply investigated in industrial processes due to the athermal processing phenomena [21]. In 1991 KLM appeared and with its simple configuration it proved that self-focusing in a laser medium in combination with an aperture behaved like a saturable absorber [31; 56]. Thus, it represented a relevant achievement of a successful fast saturable absorber of solid-state lasers and became a fundamental technology for the generation of fs pulses [57]. Another important event in 1991 was the implementation of CPA technology to Ti:sapphire laser mediums, which perform very well thermally and optically supporting ultra-short pulses (convenient near infrared broad-gain bandwidth) [25; 58; 59]. To close the historical development of technologies that contributed to achieve femtosecond pulses, in 1992 the semiconductor saturable absorber mirrors were invented (described in subchapter 2.1.2) [60; 61]. SESAMs came as a solution to the challenges of Q-switching passive mode-locking, which represented instabilities due to the modulation of a Q-switch pulse over the mode-locked pulses and its consequent constant laser off [62], thus its self-starting performance, stability, flexibility (cavity design) and reliability of mode-locked diode-pumped ultrafast solid-state lasers not only contributed to shorten pulse length but also increase the pulse energy, pulse repetition rate and its applicability out of experimental environments [29; 31; 62; 35].

As reviewed in the concept of mode-locking (chapter 2.1.2), the emission of an ultra-short pulse requires a wide beam spectrum, i.e., the bandwidth of a laser beam is inversely proportional to its pulse length; therefore, to be able to obtain shorter pulses, a laser medium with broad constant emission spectrum is required [18]. CPA technology is one approach that makes possible the shortening of the pulse and the handling of its inherent characteristics, e.g., pulse energy and pulse frequency, to avoid failures [18]. Thus, a typical configuration of a femtosecond laser for material processing consists of two important components of CPA technology: a chirped pulse amplifier CPA and a pulse compressor, like for instance, in a Ti:sapphire source [21]. These components are part of a mode-locking system as shown in subchapters Mode-locking and Supporting technologies. The process of the generation of a femtosecond pulse is initiated by mode-locking where a pulse of nanojoules energy is emitted. This domain of pulse energy (nJ) is in many cases not enough for the application and an amplification to micro or millijoules is required. This kind of direct amplification to high energy might suggest an imminent damage of the optical components and a stretching of the pulse out of the femtosecond range (longer pulse) [20]. For this reason, the implementation of CPA technology supports the pulse energy conversion to avoid the risk of damage. Three main operations are part of the CPA technology after a pulse is generated from the oscillator; first, the peak power of the pulse is weakened by making the pulse length longer by a factor of 10^3 to 10^4; then, this long pulse is amplified in a second operation, and lastly, the laser beam diameter is magnified to avoid damaging the optics and the pulse is compressed back in a compressor to regain its original characteristics of length but with an amplified energy (cp. Subchapter Supporting technologies) [18]. Figure 13 has shown before the use of gratings for dispersing the original frequency components of the pulse and making them differ in their directions and consequently in their optical path lengths (stretching by the difference in the frequency component travel time of their optical path lengths); moreover, an arrangement of parallel gratings compresses the frequency components to almost their original pulse state based on the fact that their separation distant defines the amount of compression [18; 20]. After these three operations, an ultra-short pulse is emitted in a range of energy, e.g., of mJ and a fs length.

Figure 15 illustrates a schematic configuration of a fs laser system based on a Ti:Sapphire example. It consists of two main parts: diode lasers and solid-state lasers as a part of the pumping system, and a stretcher, an amplifier and a compressor, as a part of the CPA Technology. Other typical fs lasers include mirrors or prims for dispersion compensation and a SESAM, among other possibilities [25; 59].

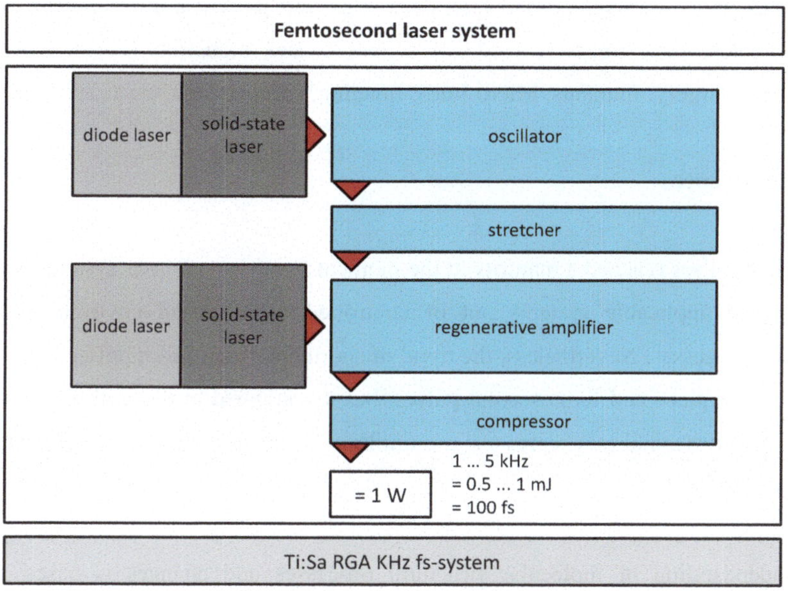

Figure 15: Schematic configuration of a fs laser system [2]

Typical active mediums used for fs lasers are mainly grouped between dyes and solid-state mediums, such as, CW Dye, CPM Dye, Ti:sapphire, Nd:glass, Nd:fiber, Er:fiber, Yb:YAG, Cr:YAG, Cr:forsterite, Cr:LiS(G)AF, among others [18; 57].

Some of the advantages of fs lasers in material processing are high intensity output, high adaptation, capability and flexibility to process several types of materials (including organic, transparent and ultra-hard), high accuracy, clean and free of recast results, ability to have a detailed processing (miniaturization) [21; 23]. The main challenges of fs laser systems are the cost of the equipment, the adaptation to industrial processes and reliability (related to its system complexity and reflected in on-time, maintenance and failures) [2]. Some of the common application fields of fs laser systems are micromachining of photovoltaics, medical devices, ultra-hard materials, polymers, among others; additionally, they are used in microstructuring, nanodrilling, microcutting and microablating to mention some industrial processes [3; 23]. The focus of this work is targeted to micromachining applications, especially, in laser beam ablation. However, the application fields of femtosecond lasers are wide and present a development trend in several fields, e.g., military, medical, scientific and industrial (Physics, Biology, Chemistry, Surgery, Implants, Microfluids, Imaging, spectroscopy, etc.) [2; 63].

Further advances

Ultra-short pulses reached a maturity at the moment femtosecond laser systems achieved reliable and applicable systems out of controlled experimental environments, e.g., industrial processes. Nevertheless, the trend of shortening laser beam pulses did not stop at the rise of pico- and femtosecond pulses, but it continued at the birth of attosecond pulses. An attosecond (as) represents a quintillionth part of a second, i.e., 1 attosecond equals to 10^{-18} s.

The understanding of molecular structural processes and changes is based on the picosecond – femtosecond timescale; however, for ultra-fast processes, such as, electronic changes, attosecond time regime is required [64; 65; 66]. The scope of the attosecond regime mainly includes the atomic distances, the ultra-fast dynamics of charges, e.g., electrons, and their relationships [64].

The first appearance of measured attosecond pulses is claimed to be in 1999 [67]. This has been possible thanks to the ionization of rare gases by a high power fs pulse that causes short-wavelength harmonics at multiples of the original fs frequency and a selection of determined harmonics, whose combination should release a train of attosecond pulses (theoretically of 100 as) [66]. The main challenges in 1999 were the generation, measurement and acknowledge of attosecond pulses [66]. In 2001 it was possible to emit light pulses at a rate of approximately 650 attosecond; these pulses were aimed to register the emission of an electron with a time resolution of $\leq$ 150 attosecond by means of a "subfemtosecond soft-X-ray pulse and few-cycle visible light pulse" (attosecond spectroscopy and metrology) [2; 65; 68]. The following years a consistency in the generation of attosecond pulses by intense light waveforms was reached; the generation of controlled 250 attosecond pulses sharing the same characteristics in the soft-X-ray regime was achieved in 2003 [68; 69]. Further research contributed to generate shorter attosecond pulses, e.g., 80 attosecond in 2008 and 67 attosecond in 2012, based on advances in nonlinear optics [70; 71] and a phase-mismatch technique [72; 73], respectively. Recent developments concentrate in the design of attosecond systems and their gain mediums; for example, the analysis and comparison of gas and solid gain mediums in the harmonics generation to shift laser light to shorter pulses by intensifying it through the gaseous or solid (a frozen rare gas) gain medium in order to increase laser power and obtain shorter wavelengths [74; 75].

High Harmonic Generation (HHG) is the main responsible for the creation of attosecond pulses. The development of ultra-short pulse lasers, e.g., Ti:Sapphire femtosecond systems, has contributed to this milestone of miniaturization of the pulse [76]. These systems allowed a focused peak intensity of more than 10^{15} W/cm^2, i.e., a high laser beam electric field, causing electronic processes in an atom to present highly nonlinear effects [76; 77]. Therefore, the focusing of a high intensity IR laser beam into a rare gas allowed its up-conversion into a coherent short pulsed extreme ultraviolet XUV beam (soft-X-ray) [76; 78]. The beaming of the IR laser into the gas creates a harmonic sprectrum or plateau having a constant amplitude followed by an exponential cutoff (high harmonic order spectrum) [76; 79]. The achievement of XUV light meant likewise the arrival of attosecond pulses, and their further development, e.g., attosecond pulse trains (APT) and isolated attosecond pulses (IAP) [23; 29; 80].

Attosecond science and attosecond lasers stand in the state-of-the-art and begin their ascendant development in their life curve. For this reason, the applications, in which they are present are mainly related to experimental environment. However, there are some practical cases in metrology, medicine and spectroscopy [73; 80; 81]. Attosecond time-scale belongs to the electronic dynamics and field of great interest, thus, molecular applications of attosecond pulses for the characterization and analysis of atomic and subatomic components are present [64]; attosecond electron wave packet interferometry, quantum path selection [79], inner shell spectrography, attoclock measurements and valence electron motion [73; 80] are examples of this application level. Practical applications include the modeling of chirurgical cornea operations [81] and the potentiality of processing material from their molecular structures thanks to the theoretical attosecond peak intensities [78; 80]. The further development of attosecond science, as some authors refer to it, might contribute positively to approach material processing at an atomic and electronic level.

2.1.4 Picosecond lasers

The development of picosecond lasers emerges from the development of mode-locking technology. In search of shorter pulses, active mode-locking was demonstrated using a helium-neon laser with an acousto-optic modulator and emitting only nanosecond pulses in 1964 but preparing the field for the following technological achievement [82; 49]. It was observed that the generation of a group of very short pulses was emitted in a ruby laser using a saturable dye (absorber) instead of a modulator in the cavity leading to the discovery of passive mode-locking in 1965 [83; 31].

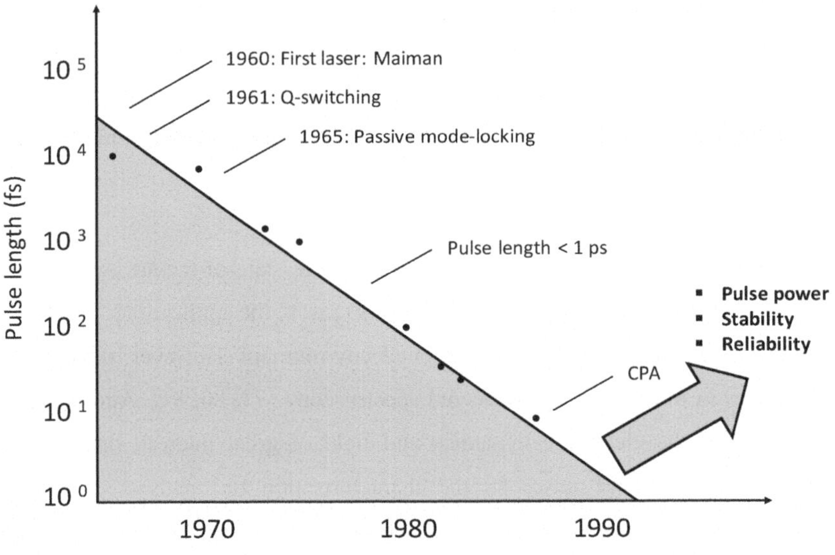

Figure 16: Development of pulse length since the invention of laser [2]

The lack of a modulation by means of saturable absorbers led to abstain from using electro-optic or acousto-optic modulators, and consequently, to avoid their shutter times in ns range. This importantly contributed to the generation of ultra-short pulses and in 1966, the first pulses in the order of picosecond range were obtained using a passively mode-locked Nd-glass laser with a broader gain profile [49; 84; 85]. The use of thin films of saturable absorber in combination with an efficient broad band response of Nd:glass and Nd:YAG permitted the generation of pulses of 10 ps length [34; 49]. These mode-locked lasers based on passive mode-locking and offering high power output made possible the birth of ultra-short pulses in picosecond range and contributed to the minimization of pulse length, as seen in subchapters mode-locking and 2.1.3 [85]. Figure 16 illustrates the development of pulse length from 1960 to 1990 including relevant historical achievements in the field.

The development of ultra-short pulse lasers followed the inertia of miniaturization of the pulse driven by technological advances. Nevertheless, reliability, stability, applicability and compactness (when compared to shorter pulse systems) of ps systems kept them contemporaneous and valid, and these characteristics accelerated its evolution. Their configuration based on mode-locking might vary from manufacturer to manufacturer and be discrete due to confidentiality, but it has a common understanding, explained in the following lines. Figure 17 illustrates a configuration of a ps laser system; it is composed of pumping sources, such as a diode sources, an oscillator, and a regenerative amplifier [2].

The comparison between this ps laser system and a fs laser system (see figure 15) contrasts the simplicity of the first one by lacking, in this case, a stretcher and compressor. In some configurations, there are pulse pickers to select one pulse or more among hundreds of them to influence the pulse repetition rate and a harmonic generator to switch from infrared IR wavelength to green and UV´s (from 1064 nm to 532 nm and 266 nm, respectively) [9].

Specifically, the configuration of a ps laser system includes in the oscillator a mode-locking component, which might be active or passive [25]. An active mode-locked ps laser system presents an acousto-optic modulator (AOM) and a passive mode-locked ps system includes a semiconductor saturable absorber mirror (SESAM); both options have to be specifically designed in the cavity to allow laser operation and to define an accurate repetition rate, among other characteristics [25].

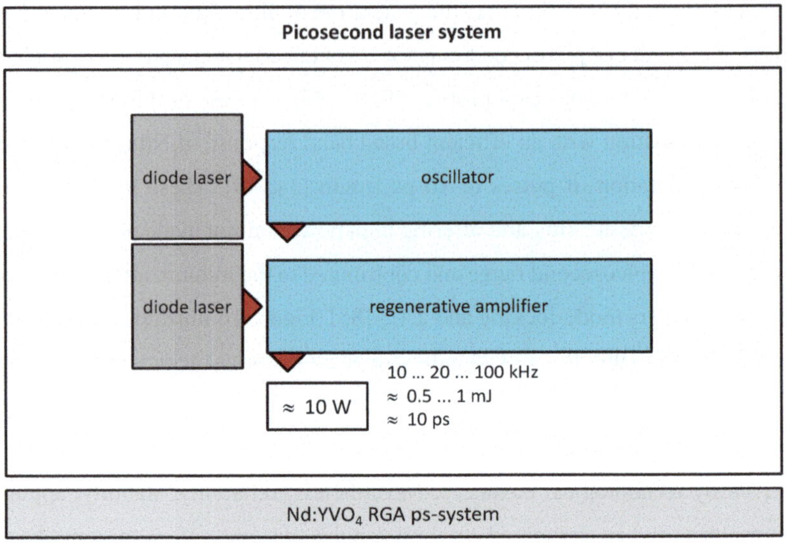

Figure 17: Schematic configuration of a ps laser system [2]

Picosecond pulses can be generated based on a solid-state [25] or a fiber system (rare-earth-doped fibers) [86]. Example of gain mediums for solid-state ps laser are Nd^{3+}:YAG and Nd^{3}:YVO_4, which work well with high power diode pumping and offer a convenient amplification bandwidth for picosecond pulse generation (e.g., from 20 to 1.5 ps) [25]. A basic picosecond material processing laser system can be constituted, additionally to the laser source, of a beam delivery system (laser scanner with F-Theta lens, beam expander, and a Z-axis-shifter), a X-Y processing plate, gases and chips extractor and off-line camera [9].

Industrial available ps laser sources offer a wide range of pulse repetition rates, e.g., from 200 to 1000 kHz [87], likewise, it has been introduced a feature related to the repetition rate named burst-mode. This mode consists in the partition of a picosecond pulse in up to ten parts in micro pulses or bursts. These groups of bursts divide the original pulse energy into the number of them and have a nanosecond interval between each of them, as shown with 8 bursts in figure 18 [88]. Every group of bursts is released accordingly to the configured pulse repetition rate as in single pulse (no burst) mode.

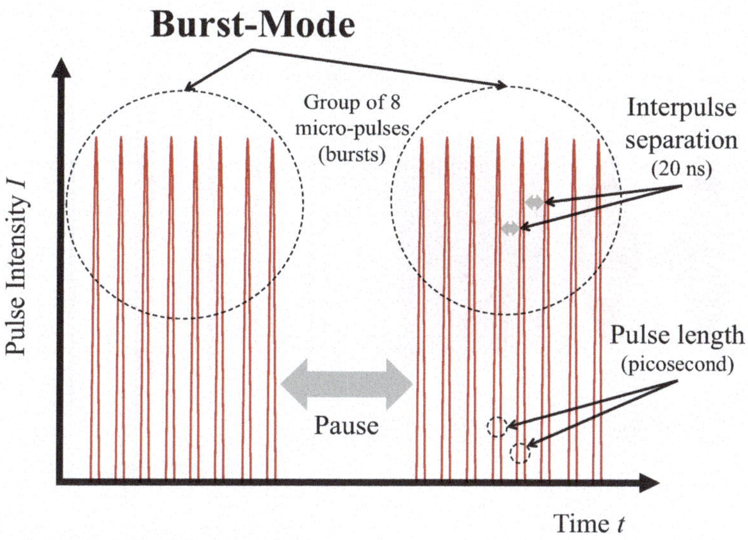

Figure 18: Schematic representation of burst-mode with 8 bursts [88]

The burst-mode technology platform uses an oscillator of < 10 ps that emits a pulsed beam (at approximately 50 MHz) travelling through a switch and arriving to a transient amplifier [9; 89]. The aperture of the amplifier takes longer to close, and this permits the creation of laser bursts of an approximately 20 ns separation [89]. The idea of the bursts-mode technology platform is illustrated in figure 19.

There are different approaches to obtain bursts: optical set-ups and in-resonator set-ups, based on AOM/EOM and amplifiers [90]. The performance of burst-mode in laser processing has been evaluated in the last years and continues to be of great interest for further research [88; 90; 91; 92].

The application of picosecond lasers in material processing has been achieved thanks to their convenient technological characteristics and advantages. Picosecond lasers have the ability to compete versus sources with a shorter pulse length, like for instance, fs lasers. They are able to achieve similar qualitative and quantitative ablative results with metals, hard-materials (synthetic diamonds and tungsten carbide) and semiconductors, e.g., silicium for solar panels, and benefitting from their high pulse repetition rates.

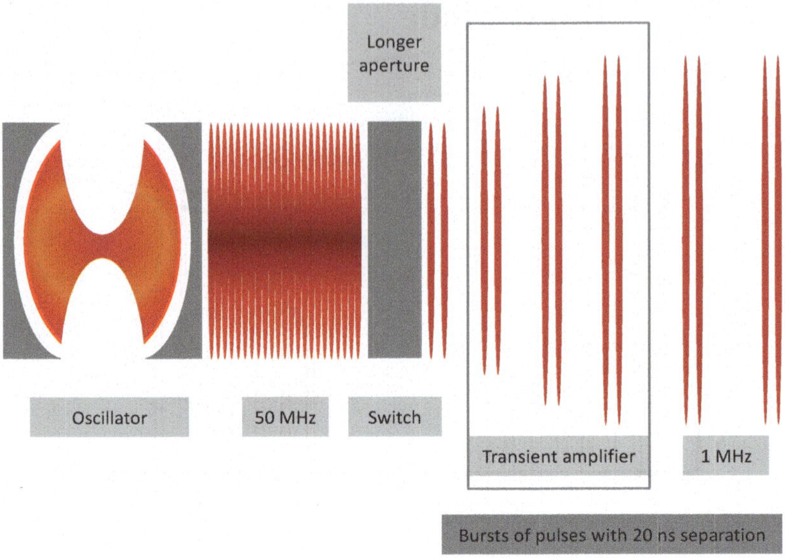

Figure 19: Burst-mode technology platform [89]

In addition, they are able to perform high average power levels which contribute to obtain an almost "cold" ablation [2]. Their systems are robust, compact, less complex, flexible and reliable because they do not suffer high damage risk of their optical components due to high power density (in contrast to femtosecond pulse lengths), do not require a specific-wavelength pumping source, and can be, in some cases, diode pumped directly to the gain medium (abstaining from elements like seed laser chirping and compressors) [2; 93]. In general, ps lasers offer the possibility of an economic activity by maximizing the productivity of certain applications and materials. Their reliability and technological capabilities make them suitable to be used in different application fields over the last decade; for example, research and development, mechanical engineering, nano manufacturing, electronics, photovoltaics, life science and metrology [4]. Some examples of applications of picosecond lasers are described below.

Table 2 offers a wide perspective of the application possibilities of ps lasers from medicine to manufacturing. This kind of pulses enables processes like laser beam cutting, laser beam ablation, laser beam structuring, laser beam drilling, laser beam welding, laser beam marking, laser beam scribing and laser beam internal engraving, among others.

Table 2: Applications of ps laser

Applications	Remarks	Source
Micro processing, laser beam structuring and laser beam ablation	Metals and other suitable materials (circular forms in silicium and quartz glass). Reparation of lithography masks. Fabrication of photonic elements by deep laser processing.	[2]
Laser beam cutting and laser beam ablation in Medicine and implantology	Surgery of organic tissue by means of laser cutting and ablation in ophthalmology, otolaryngologist, odontology, neurology and cardiology. Surface structuring of implants and production of micro stents.	[2] [4] [22]
Functional laser structuring	To achieve surface with functional characteristics, e.g., repellent, adhesive, low friction, anti-reflective, self-cleaning. Riblet production for aerodynamics.	[3] [4] [94] [95]
Laser beam ablation and micro drilling	Fabrication of embossing tools and injection nozzles (WC, steel, etc.), transparent materials processing (glass), processing of ceramics and polymers, thin film processing in photovoltaics, texturing, laser tooling (micro mold). Process modeling.	[3] [4] [96] [9]
Laser beam cutting and laser beam ablation in solar engineering systems	Ablation of layers for efficient energy transfer in photovoltaic materials; production of perforations, isolation of edges and cut of plates.	[93]
Welding and marking	Micro welding (fusion welding), internal engraving and scribing of transparent materials	[9]
Laser beam processing of hard materials	Cutting of ultra-hard materials (PCD, CBN, WC, sapphire), edge rounding on high-precision carbide tools, marking of diamond. Processing of monocrystalline diamond MCD, Chemical vapour deposited diamond CVD-D, PCD and polycrystalline cubic boron nitride PCBN. Fabrication of cutting inserts (contour, negative chamfer and chipbreaker). Dressing tool preparation, preparation of cutting edges, profiling, cleaning, sharpening, structuring and touch dressing of diamond-based grinding wheels.	[9] [97] [98]

Moreover, they are capable of processing different kinds of materials; for example, plastics, ceramics, metals, organic matter, compound materials, and ultra-hard materials. Thus, ps laser beam ablation of ultra-hard materials represents an important subject of research of this thesis and an area of high interest of the manufacturing industry.

2.2 Laser ablation

As an important part of the background of this work, subchapter 2.2 belongs to the first phase *plan* of the systematic quality-oriented approach (see figure 4). It refers to the concept and fundamentals of laser beam ablation of material and sets a basis for the processing of ultra-hard materials. Laser beam ablation requires several system parameters (laser source, processing strategy, system parameters and material properties) to achieve the objectives of productivity, i.e., ablation rate, and quality, e.g., surface quality, thus it is important to understand the influence of the laser source parameters on laser beam ablation of materials.

2.2.1 Definition

Laser beam ablation, from a pragmatic-oriented definition [15], is the finishing process of material removal in solid, liquid or vaporous state from a solid body by heat-induced processes. The required heat to ablate a material is generated by the effective energy released from a laser beam source. Moreover, laser ablation denominated as "cold ablation" is defined as the laser beam sublimation (ablation) and is the separation of material of a solid body from solid to vaporous state without a liquid phase by heat-induced (laser) processes [15].

For this thesis, picosecond laser beam ablation is a pyrolytic process on the material that consists of the heating, melting and evaporating of molecules of a solid by means of optical energy induced processes [14; 15; 16]. As seen in subchapter 2.1, especially in figure 5, picosecond pulses, like the pulses applied in this investigation τ_H = 10 ps, belong to the ultra-short pulses, where the diffusion depth d is in the same magnitude or less that the optical penetration depth δ during a pulse interaction. Therefore, after the absorption of the laser beam, the ablating process starts; this can be, depending on its pulse length, photo-induced or thermal ablation, offering the first, photo-induced, more ablation precision than the thermal (processing diameter / focal diameter) [24]. Figure 20 illustrates the boundaries of photo-induced and thermal ablation, specifying the times scales, the threshold fluence F_{th}, and the input (applied) fluence F_0 to obtain "cool, hot or melt ablation".

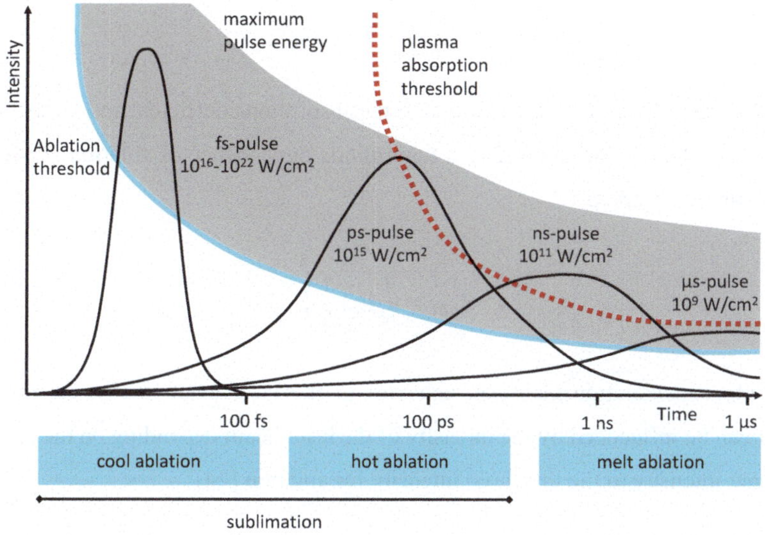

Figure 20: Time scales and processes for material ablation [4]

2.2.2 Thermal and photo-induced considerations for laser beam ablation

According to Poprawe [24], the interaction of laser beam intensity, absorption properties of the materials and the processing interaction time dictates the ablation results on the workpiece and thus, various influencing parameters have to be taken into consideration. One of them are the material properties such as reflectivity, absorptivity, melting and evaporation latent heat. Parameters to be considered in the selection of the laser beam source and optics are laser beam radius ω_0, pulse length τ_H and wavelength λ. The absorption of the laser radiation might be linear or non-linear and depends on the characteristics of the laser beam and material. Linear absorption, absorption not depending on laser intensity, is present on a material, when the laser wavelength belongs to the absorption band of that material or when the material has due free or quasi-free electrons, like in metallic materials [24]. **Beer-Lambert law** describes linear absorption, and states that any electromagnetic radiation, such as laser radiation, travelling through a material, is reduced exponentially with the thickness for a given wavelength [24; 99]

$$I(x) = (1 - R)I_0 e^{-\propto x} \qquad (4)$$

Where, R is the reflectivity of the material, α the absorption coefficient and I_0 the threshold intensity for ablation. Based on [24], a logarithmic dependence of ablation depth Z_{abl} on intensity can be observed

$$Z_{abl} = \frac{1}{\propto} \ln\left(\frac{I}{I_0}\right) \qquad (5)$$

Taking into consideration Gaussian behavior of the laser beam focus, the processing diameter can be influenced by the intensity of the laser beam depending on the proportion of the input intensity to the threshold intensity for ablation [24].

Two-temperature model

The ultra-short pulse regime, i.e., picosecond regime, has shown a totally different behavior to laser beam ablation when compared to longer pulses. Less HAZ and higher accuracy are a result of the laser-material interaction under these ultra-short pulse regimes. At the free and bound electrons of the substance, the electromagnetic energy is absorbed, and there, the photons interact with the material and heat the electrons, which transfer their energy to the lattice of the material after a determined time [3; 24]. This laser-matter phenomenon is described by the **two-temperature model TTM** stating that it happens among photons and electrons, having an overheated electron gas, transferring energy to the lattice, and depending on a material specific coupling coefficient [4; 24]. The schematic representation of this laser-matter interaction is presented in figure 21.

Pulse length influences the effect of absorbed laser radiation on thermal ablation of a material. Furthermore, the absorption effects, behaving linearly, are subdivided in four regimes. The first describes the absorption of the laser optical energy by quasi-free electron (e.g., copper, $t_{\gamma e} < 10$ fs), the second explains the thermalization of the electrons (electron system, $t_{ee} < 100$ fs), the third describes the interaction of the electron and phonon system ($t_{ep} < 10$ ps), and the last regime presents the thermalization of the phonon system ($t_{pp} < 100$ ps) [24].

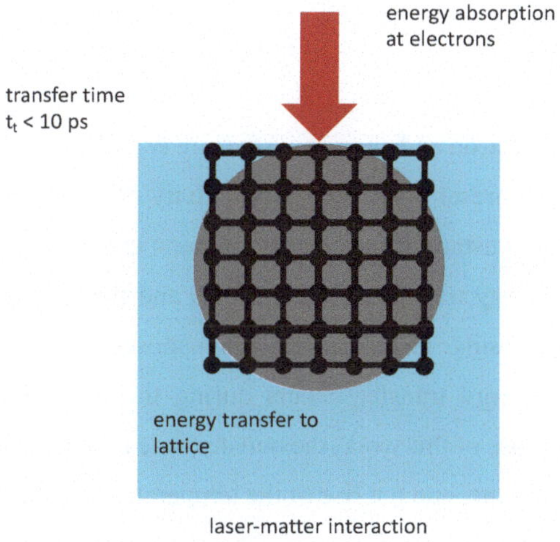

Figure 21: Principle of laser beam interaction with materials [3; 24]

Hence, it is possible to have either pulse lengths longer than the thermalization of phonons ($\tau_H > t_{pp}$), or pulse lengths shorter than the electron-phonon relaxation time ($\tau_H < t_{ep}$). The first case neglects the periods for absorption of the photons, thermalization of electron and phonon interaction and forces the system to behave as a one temperature model for the electron and phonon system [2; 16; 24]. Laser ablation behaves as a simplified and immediate process. The description of thermal diffusion depth is described in terms of thermal conductivity of the material k, heat capacity c_p (specific heat), density ρ and pulse length of the emitted radiation t_p

$$d = 2\sqrt{\frac{kt_p}{c_p\rho}} \qquad (6)$$

The second case, $\tau_H < t_{ep}$, behaves as decoupled electron and phonon processes and follows the description of the two-temperature model in two coupled differential equations

$$c_e \frac{\mathrm{d}T_e}{\mathrm{dt}} = \frac{\partial}{\partial_\zeta}\left(k_e \frac{\partial T_e}{\partial \zeta}\right) + S - \mu(T_e - T_p) \tag{7}$$

$$c_p \frac{\mathrm{d}T_p}{\mathrm{dt}} = \mu(T_e - T_p) \tag{8}$$

The heat capacities of the electron and phonon system are c_e and c_p, and their temperatures T_e and T_p, respectively; k represents the heat conductivity of the electron system, S the input optical energy, μ the constant of the electron-phonon coupling. Like shown before, figure 21 illustrates the energy transfer of the systems and the way normally occurs in picosecond laser beam processing. Thus, it contrasts to nanosecond laser beam processing, for example, where the energy transfer occurs during the development of the input nanosecond pulse. For the case of this work, the transfer time in metals, like WC, normally occurs in the picosecond regime, and the rise of the temperature occurs after the pulse has been irradiated. Moreover, the use of a picosecond laser beam source not only contributes to avoid thermal side-effects on the material but also to obtain a precise ablation quality thanks to its high intensities (10^{10} W/cm^2). These levels of peak intensities reduce the melting phase of the material and increase the vaporization of it. Also, these peak intensities are able to sublimate almost all materials [24] and help to design ablation system to obtain relevant results in ablation rates and surface quality in a wide range of materials.

Incubation

One last relevant aspect to be considered in laser beam ablation, due to its pulsed nature, is the **incubation effect**. It consists in the effect of multiple pulses entering a material one after another in a same spot or area in a determined rate, energy and time, and how it changes the material properties. Thus, one important aspect corresponds to how the threshold fluence is affected by a material-related factor [100; 101; 102]. The following lines describe this phenomenon.

An early mention of an incubation factor or accumulation phenomenon appeared in 1987 in a doctoral thesis on multiple pulse laser induced damage on metals describing a threshold fluence reduction in accumulation by a multi-pulse irradiation of single crystal metals surfaces at an infrared beam [103]. It is explained that "the storage cycle of thermal stress-strain energy induced by a laser pulse" might be responsible of the accumulation behavior and consequently of a decrease of threshold fluence on the material [102]. Based on that and on other studies [101; 104; 105], *incubation* can be defined in the event of ultra-short pulse laser beam processing as the modification of the properties of the material and its threshold fluence by a material-related factor S due to multiple pulses irradiation on a same spot of the surface and affected by each subsequent pulse from 2 to N pulses. An analysis of the development of threshold fluence in function of the logarithmic number of pulses was reinterpreted in terms of accumulations curves. These investigations of cumulative effects [103] released several accumulation curves illustrating the behavior of the product of applied number of pulses and threshold fluence versus the number of applied pulses and plotted as log. This contributed to understand the cumulative effect (incubation) from the linear section of the curves (slope of the graph) and to foresee its impact as expressed in equation 9:

$$F_{th(N)} = F_{th(1)} \cdot N^{S-1} \tag{9}$$

Where, $F_{th(N)}$ corresponds to the threshold fluence of N pulses, $F_{th(1)}$ is the threshold fluence of one pulse, N represents the number of applied pulses and S the incubation factor, as mentioned before.

Figure 22 schematically illustrates the phenomenon of incubation (A) and an example of accumulation curves for copper [102] (B).

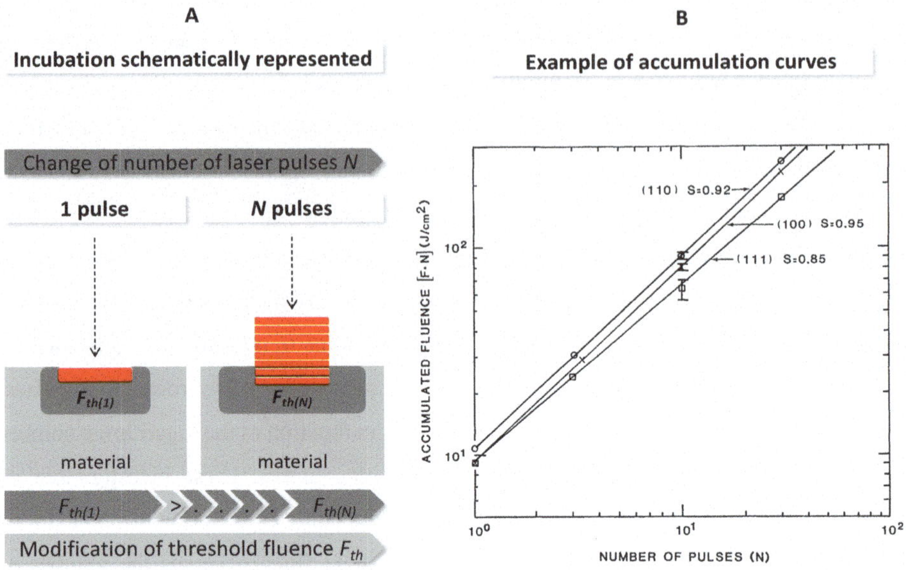

Figure 22: (A) Schematic representation of incubation and (B) an example of accumulation curves for copper Cu (three different crystal orientations) from Jee in "*Laser-induced damage on single-crystal metal surfaces*" [102]. The slope of the curves represents the incubation factor *S*

It can be observed that the irradiation of pulses in the same area modifies the F_{th} of the material and influences the laser beam ablation process as a function of the number of pulses. The quantification of the data by the logarithmic plot of the accumulated threshold fluence versus the number of pulses describes the material-related *S* factor.

The incubation factor defines whether the influence of incubation is present or not in the material and it is described by $S = 1$ when no incubation is present; *S* from 0 to 1 when the material is affected and its F_{th} is lower; and *S* greater than 1 when the material is influenced in the opposite direction, i.e., its F_{th} is higher [100; 101; 102; 106]. The identification of the *S* factor for this work is the result of an experimental and analytical methodology on cemented WC that is executed in the first part of chapter 5.

2.3 Cemented tungsten carbide

The following lines present a description of cemented tungsten carbide, its historic development, production, characteristics and advantages as its relevance as the chosen material for this work.

In the universe of hard materials, the heavy metal carbides' group, also known as *hardmetal*, sintered carbide or cemented carbide, represents an important intermetallic compound that has been historically preferred for cutting tools, wear-oriented applications and machining thanks to its high hardness, melting point, availability and tenacity, among others characteristics [107; 108]. Moreover, a heavy metal carbide that has been successfully used in the cutting and wear application fields thanks to its hardness, stiffness and high-temperature oxidation resistance is tungsten carbide WC [109]. The choice for tungsten carbide has prevailed in the market, when the application allows it, among other materials thanks to its properties even though high technological alternatives, such as polycrystalline diamond PCD and cubic boron nitride cBN, have been developed and complemented the applications fields [109].

The origin of tungsten carbide has its antecedents in Germany in the development of an alternative for wire drawing dies made of diamond that were used for tungsten filaments in the first decades of the last century [110]. Thus, the sintering of tungsten carbide and cobalt (Co) not only represented a success in the production of drawing dies because of its hardness and wear capabilities but also started an incursion in the production of cutting tools, wear components and machine elements [110; 111]. By the end of the 1930's, cemented tungsten carbide, with cobalt as a binding agent (6%), was introduced to the market by the company Krupp as an alternative to diamond and under the name *Widia* [110; 112]. Figure 23, extracted from Fukuhara [111], illustrates the introduction of diverse cutting tool materials through time as a function of its cutting speed; there it can be observed the early appearance of sintered carbides in the modern history.

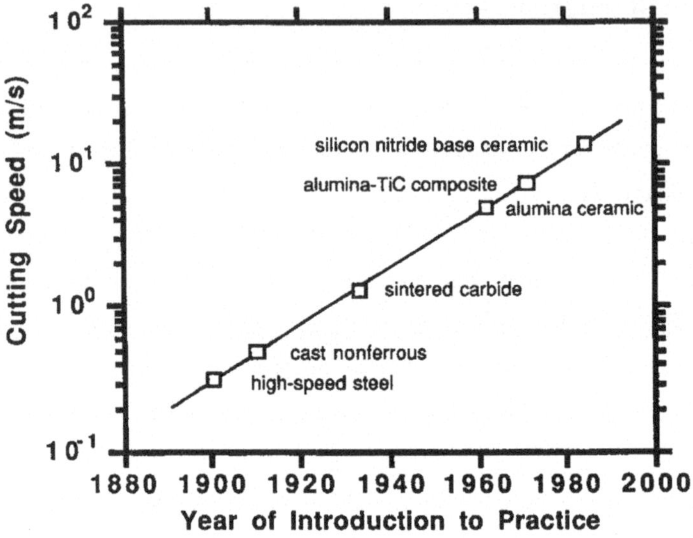

Figure 23: Improvement of tool materials [111]

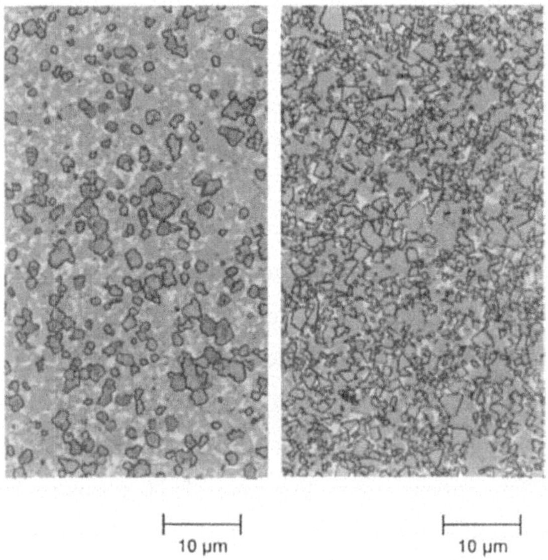

Figure 24: Microstructure (light-optical microscope) of a WC/Co-based hardmetal (WIDIA TTR) containing hexagonal WC and cubic carbides (right: cubic phase, left hexagonal phase enhanced by specific etching) [113]

Tungsten carbide is formed by a combination of WC micrograins and a binder matrix that supplies tenacity for the bond (lower melting degree metal from iron group), normally of cobalt [20; 107]. The carbide grade, i.e., hardness of the compound, is mainly defined by the following descriptive parameters: The content percentage of the mixture of cemented tungsten carbide that can be from 94% WC and 6% Co to 88% WC and 12% Co and the size of the grain that can present micrograins (particles under 1 μm) and submicrograins (particles under 0.5 μm) [20; 107]. Figure 24 presents an image of a microstructure of a WC-Co sample under a light-optical microscope (taken by Lengauer in [113]).

The hardness of cemented WC increases, e.g., with lower cobalt content and smaller tungsten carbide grain size [20]. The crystal structure of WC presents a simple structure with a hexagonal cell (a = 290pm, c = 284 pm) and isolated carbon atoms within the trigonal W prism, as illustrated by Jeitschko (see figure 25) [107; 108; 110].

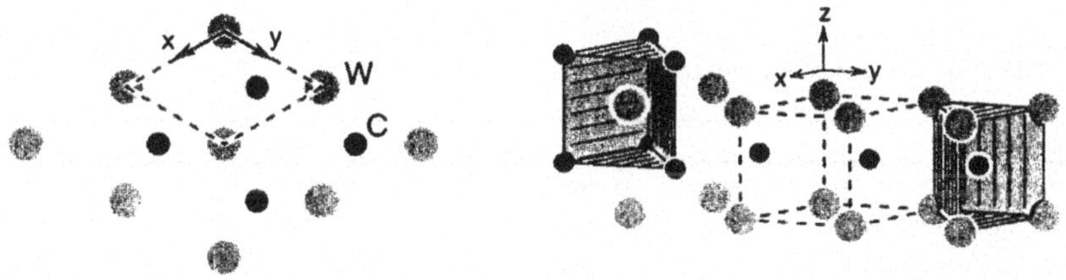

Figure 25: Crystal structure of WC. A projection along the hexagonal c axis is shown on the left-hand side of the drawing and a view of the unit cell is presented on the right-hand side [107]

WC-Co heavy metal carbides, i.e., cemented tungsten carbide, are compounded by tungsten monocarbide (WC), an elemental component that is produced from the carburization of tungsten; the latter is mainly obtained from the minerals Scheelite and Wolframite [110]. The production of cemented WC is done under the guidelines of powder metallurgy and fundamentally described by Cardarelli [114]. A carburization process generates a tungsten carbide powder that is blended with the binder, such as cobalt. Afterwards, a consolidation of the powder into net compacts and billets by pressing and extruding is achieved. The resulting product can be machined and presintered for shape and handling. The process ends with sintering and post-sintering of the compound.

An example of a traditional manufacturing process of cemented tungsten carbide end-products, such as cutting inserts, is grinding with diamond and cubic boron abrasives (cBN) [115]. The challenges of using diamond and cBN, among others, for producing hard materials allow ultra-short pulse lasers to target and develop this field.

The maturity of tungsten carbide has allowed the material to be a constant solution in tool, cutting, wire drawing die, drill and wear application fields thanks to its advantages (properties) [113]. Examples of them are machining thermal and chemical stability, high mechanical strength, higher hardness than steel, high wear and abrasion resistance, fast heat transfer, and appropriate for environments presenting high wear, abrasion, corrosion and shock [20; 114].

A summary of mechanical, thermal and electrical properties and characteristics of tungsten carbide is presented in table 3.

Table 3: Material properties of tungsten carbide

Properties	Value	Unit
Thermal conductivity k	164	W m^{-1} K^{-1}
Hardeness (Vickers) HV	1700 – 2400	HK
Young's modulus EM	668	GPa
Tensile strength TS	344	MPa
Compressive strength CS	2683 – 2958	MPa
Linear thermal expansion coefficient CTE	3.9	10^{-6} K^{-1}
Electrical resistivity ER	18	μΩ cm
Specific heat (solid-state) $C_{p,s}$	210	J kg^{-1} K^{-1}
Mass density (solid-state) ρ_s	15630 – 15930	kg m^{-3}
Melting point T_m	3143	K
Boiling point T_b	6273	K
Latent heat of fusion LH_f	400 000	J kg^{-1}
Latent heat of evaporation LH_e	4 020 000	J kg^{-1}

Sources: [116; 117; 110; 118; 119; 120; 121; 114; 20]

The use of this properties and characteristics and the identification and determination of others not available in the literature represent an important objective in the understanding of laser beam ablation of cemented tungsten carbide further in this work.

2.4 Quality tools

The importance of a global quality philosophy in an organization from strategic decisions to operations affects directly its survival, i.e., the management of productive resources and operations results vital to maintain a presence in the market and competitiveness [12]. In laser engineering (among others), changing processes, new systems and novel materials are just a constant event. Therefore, the integration of quality activities, efficiency, effectiveness and value assures the adaptability to change [12]. In laser processing, change is present in every aspect of the process; for example, the development of laser sources in beam power, pulse length and pulse frequency is improving constantly. Moreover, the processed materials represent a wide range of possibilities that goes from polymers and ceramics to metals and organic tissues, among others. Furthermore, the possibilities within one type of material might represent a combination of more materials in many degrees and amounts. Example of that is the tungsten carbide WC used on cutting tools, a compound of tungsten and carbon atoms positioned on a layer of Cobalt having different grain sizes and composition percentages [8]. Thus, it is required to apply a procedure that supports changes, gains advantage in time and cost, and collects a high amount of knowledge.

As mentioned before, the goal of the cycle is to obtain continuous improvement by mastering its four phases *plan*, *do*, *check*, and *act*, which are based on the Deming cycle philosophy (PDCA). Starting with *plan*, the first phase sets all the required information of the aspects of the process to foresee the operation. *Do*, second phase, executes the plans and represents the practical work, i.e., the laser beam processing of the material. The third phase, *check*, evaluates the executed actions and their relationship to the planned activities and releases a report of the status. Finally, the last phase, *act*, compares the final results with the objectives, adjusts the actions to support the learning curve and generates continuous improvement by introducing active feedback to the process.

The application of quality-oriented tools and theoretical analyses under a scientific criterion builds an empirical-analytical strategy that saves resources from inefficient research efforts, adds value and uses resources efficiently and effectively in the experimental and implemental actions.

2.4.1 Background

The *renaissance* of the philosophy of quality in the 1980's by, e.g., Juran, Deming and Crosby, brought many useful tools and instruments to help to include value in the processes and products [12]. These tools contributed to reduce variability in the process, an aspect observed as a cause of failures, and were mainly based on statistical analysis. Prevention and general management were also concepts, which were included in all areas of the organization to acquire quality in all the process chain. Based on *kaizen* or continuous improvement the PDCA cycle has become a part of six sigma, a philosophy seeking the elimination of defects in the processes and products and the minimization of variation (i.e., six sigma) from a total perspective in the organization and including all their elements when including lean manufacturing (lean six sigma) [12; 13; 122; 123]. The PDCA cycle gives great importance to the influence of the input and the lecture of the outputs; thus, hypothesis testing and DoE`s represent relevant tools to include in the analysis [12; 13]. Specific quality tools for continuous improvement and six sigma are flowcharts, run charts, pareto charts, check-sheets, cause-and-effect diagrams (Ishikawa diagram), opportunity flow diagrams, quality function deployment QFD, control charts, failure mode and effects analysis and DoE's (factorial, variable search, etc.) [12; 13; 124]. For the conditions of this investigation, cause-and-effect diagram and DoE's, like variable search, were selected.

2.4.2 Cause-and-effect diagram

The cause-and-effect diagram is also known as the fishbone diagram or the Ishikawa diagram in honor to its creator Kaoru Ishikawa in 1968; it presents a visual organization of possible relationships among potential causes and effects from particular interactions to a general objective or challenge [12; 125]. Starting from a general view, the diagram is built in a way that the causes of the objective or challenge become the objective or challenge and their particular causes have to be defined until all the possible particular and general causes of the system are included. Thus, this tool helps to have, first, a wide perspective of the objective or challenge and all their influencing factors and, two, an analysis of the causes that are influencing the system [12].

In laser beam ablation, an initial step of the investigation is the determination of the related parameters in the system and the cause-and-effect diagram contributes as a top down approach to concentrate all possible parameters that are related to the process and its goals, and presents a clear target oriented arrangement of parameter where their relationships can be schematically analyzed and following a flow that finishes at the general effect or goal [13; 88; 125; 126]. All causes that might make an effect in the ablation process are considered and taken to its immediate effect that might be a cause for other effect or goal.

Figure 26 shows the cause-and-effect diagram for laser beam ablation. It illustrates the analysis of the laser beam ablation in terms of the maximization of the ablation rate.

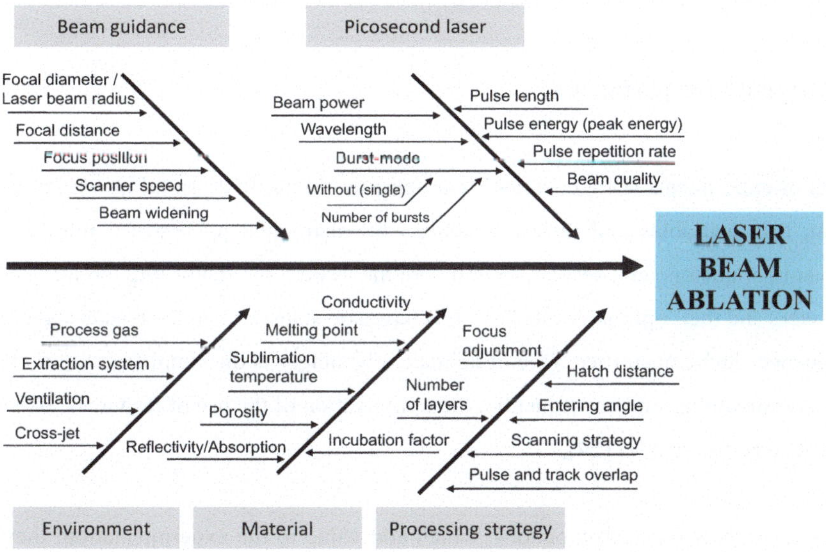

Figure 26: Ishikawa diagram for the analysis of laser beam ablation [88]

It can be observed that the diagram comprises groups of variables that might have an influence on the target and that can be either controllable and have a direct interaction on the objective (e.g., laser beam power, pulse frequency, etc.), or not controllable and linked to procedural or physical circumstances, e.g., the properties of a material. Thus, an important perspective can be obtained and a delimitation of the system behavior onto the input-output relationship can be reached.

The diagram illustrates laser beam ablation as the most important effect of the diagram and measured in terms of ablation rate. Moreover, five groups of causes that might influence the system can be observed (beam guidance, laser source, environmental aspects, material and processing strategy).

Selecting some of the parameters and neglecting others, the analysis moves to a more detailed investigation where each of the selected parameters and their interactions among them is investigated and quantified. For this kind of study, the understanding of design of experiments is required.

2.4.3 Design of experiments

Design of experiments (DoE) is a tool that helps to obtain, first, useful information of a system, their variables and their interactions by distinguishing between relevant and irrelevant parameters, and second, to optimize the system by identifying the influencing parameters and their optimal levels [13]. It is based on statistics and the results and degree of influence can be quantified. One of its main advantages is the simplification of efforts in the investigations, and consequently, the optimization of the use of resources, like time and costs (see figure 27) [124].

The DoE presents four sub-functions, which add value to the experimentation; they are system analysis, strategy of experiment, conduction of experiment and evaluation of experiment [13].

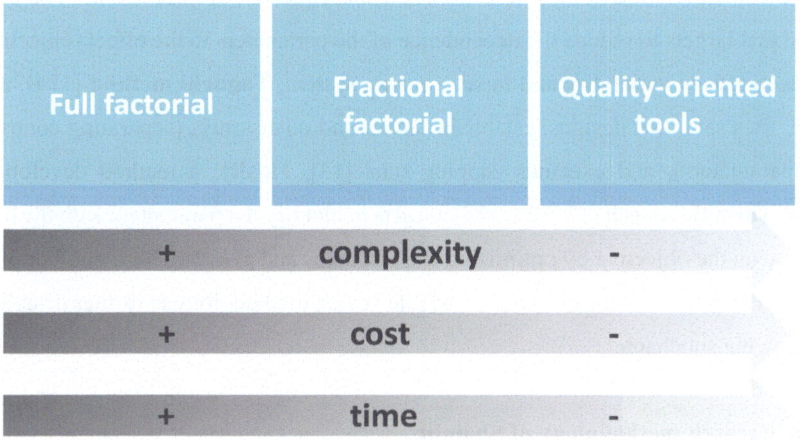

Figure 27: Advantages of DoE

Classic methods and others

There are several types of DoE that can be used for specific conditions and objectives. In the following lines, the most common designs are going to be presented. The simplest of the methods is **one-factor-at-a-time method** [13; 124], in which, as its name indicates, one variable is tested at different levels while all the others are kept constant. Even though this method is simple and easy to apply, it brings partial information due to the low number of parameters tested or high cost if all parameters are tested, also it neglects the interactions of the parameters, since only one parameter is tested. **Full factorial designs** [13] offer high degree of information since all parameters are tested and varied at different levels in a balance way. In this method main effects and interaction can be quantified and aspects like regression analysis, confidence interval and analysis of variance can be applied. Unfortunately, full factorial design requires a high number of runs and the higher the number of parameters the higher the required experimental combinations and costs. Therefore, **fractional factorial designs** [13] are used to avoid high number of experiments by selecting orthogonal arrays of parameters, which are carefully chosen based on the idea of big influence and the knowledge of their interactions. Thus, the number of runs is reduced and the focus on relevant effects is target. Nevertheless, this method presents the risk of misinterpretation of the interaction, and consequently, of false results. There is one method that targets to reduce the dependence of the parameters to the effect (objective), it pursues to reduce variability and to set a robust system; **Taguchi method** [124], as it is known, uses saturated designs, establishes inner and outer arrays (separating control and noise parameters), and executes capping runs [13]. Finally, a method developed by Shainin, **variable search** [13; 124; 127], targets to identify the parameters with the biggest influence on the objective by optimizing the resources and avoiding high number of runs. Thanks to its relevance for this work, variable search methodology is further described in the following subchapter.

Variable search methodology of Shainin

The variable search by Shainin is a DoE method that focuses on the identification of the most significant system variables based on a parameter comparison and the 20/80 Pareto principle. The objective of the method is to discard non-significant parameters and identify and confirm the most influential parameters on a selected objective. The Shainin methodology is understood as component search or variable search depending on the perspective of the analysis [127]. The concept behind both perspectives leads to the quantification of the degree of influence of significant parameters in a process. Systems and processes possessing a high number of influencing parameters, e.g., material processing and manufacturing, can reduce the number of parameters by identifying the parameters with the highest influence on the response [13]. Thus, a small, manageable and controllable number of parameters can be extensively analyzed and optimized. The nomenclature of the methodology [127] defines as "Red X" the parameters that have high influence on the output or response. The "Pink X" factors represent the parameters that share an influence on the response but in a lower degree than the "Red X" parameters.

The variable search methodology consists of three main phases, as illustrated in figure 28: Choice of the parameters and settings (levels), preliminary experimentation and main experimentation; a complementing validation phase can be added at the end of the experimentations for reliability [126; 127; 128]. After the objectives have been determined (e.g. maximization of ablation rate, reduction of defects, etc.), the first phase, choice of the parameters and settings, determines the parameters that are selected for the investigation and can be directly controlled in the system. The selection of the levels or settings has to be set for each parameter and attempting to achieve a positive qualitative response and a negative qualitative response by the selection of a positive qualitative setting and a negative qualitative setting, respectively. An example of Bhote in a metal forming operation by the use of a press break [127] defines the objective of the variable search by reaching the tolerance of the produced parts under 0.005" in a system of several parameters, such as, punch and die alignment, metal thickness and metal hardness. According to this objective, the levels of the parameters were defined as a positive setting when the level was probable to comply with the tolerance. On the other hand, the levels that were defined as a negative setting represented a level that was probable to be out of tolerance. Thus, Bhote defines for the first parameter, punch and die alignment, a positive setting as "aligned" and a negative setting as "not aligned".

The second phase, preliminary experimentation, consists basically in the execution of two experimental runs per setting with the objective of yielding reference values for comparison and setting the highest and lowest limits of the response. The first experimental runs set all parameters at their positive qualitative setting; the second runs, in contrast, set all parameters at their negative qualitative setting. The average of both experimental runs per setting defines the positive reference value V_+ and the negative reference value V_-, respectively. In order to be valid, the results of the preliminary experiments have to fulfill the statistical test D_s / $d_s \geq 5/1$ based on the classical F table. Pfeifer [13] describes the estimation of D_s as the "total effect due to simultaneous variation" and d_s as the "dispersion between repeat repetitions"(equations 10 and 11, respectively):

$$D_s = \frac{B_1 + B_2}{2} - \frac{W_1 + W_2}{2} \qquad (10)$$

$$d_s = \frac{B_1 - B_2}{2} + \frac{W_1 - W_2}{2} \qquad (11)$$

Where, B_1 and B_2 represent the results of the first and second preliminary experimental runs with all parameters set to their positive settings, and W_1 and W_2 represent the results of the first and second preliminary experimental runs with all parameters set to their negative setting.

The fulfillment of the Test D_s / d_s assures the presence of at least one dominant parameter in the system, i.e., the presence of a RED X parameter [13; 127].

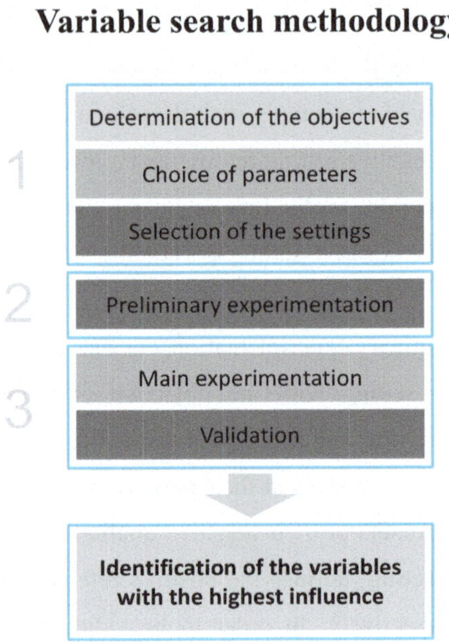

Figure 28: Phases of the variable search methodology

The third phase, main experimentation, focuses in the examination of each parameter against the rest of them by testing each of its settings [13]. Expressed differently, a parameter is set at its negative setting and the rest of the parameters at their positive setting in an experimental run. In the following experimental run, the parameter under investigation is set at its positive setting and the rest of the parameters at their negative setting. Thus, each parameter is examined following this testing procedure. The following figure 29 exemplifies the three phases of the methodology, illustrates the combination of parameters and settings, and presents the results of the experimental runs (x_1, x_2, etc.) and their corresponding setting switch (y_1, y_2, etc.) in the main experimentation.

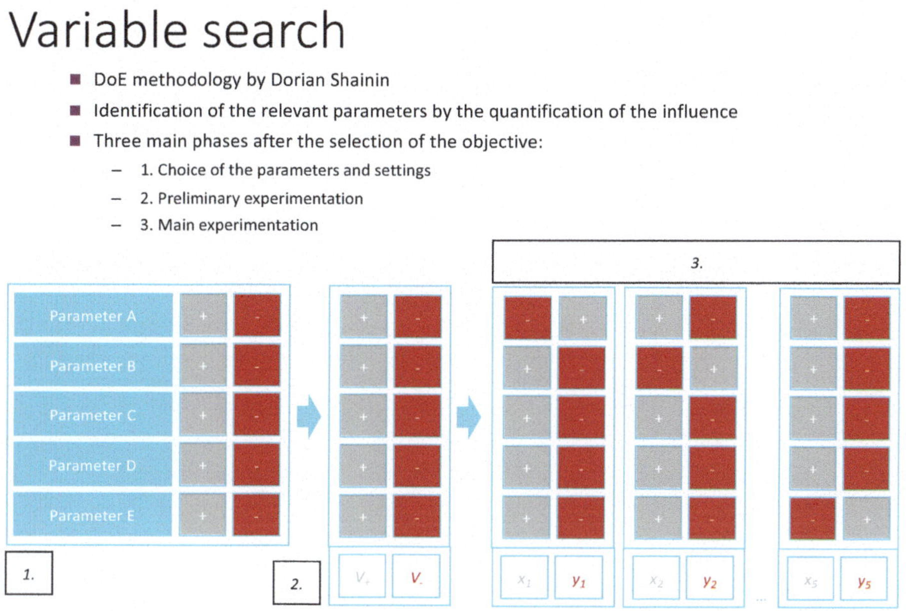

Figure 29: Variable search methodology

The execution of the main experimentation concludes with the analysis of the results. The variation of the results with their respective reference values is quantified, and thus, the parameters with the highest degree of influence on the response are found [13; 124; 127]. Once completed the methodology, a validation of the results can be done, and a detailed full factorial experimentation for further information can be executed.

3 Research concept and approaches

The development of this work is based on the philosophy of quality and the efficient application of its tools; one of its objectives is the creation and implementation of a systematic quality-oriented approach for the analysis of laser beam ablation and based on the Deming cycle (illustrated in chapter 1, in figure 2). As a part of the state of the art of this thesis, under the phase *plan* in the structure (see figure 4), the quality tools are responsible to assure a quality-oriented process, that is efficient and reliable, and adaptable to changing environments, like new materials and systems (flexible). Based on the phases, this work pursues the systematic identification, description and quantification of significant influencing parameters on the ps laser beam ablation of WC, targeting ablation (e.g., ablated volume, ablation depth and ablation rate). Thus, a systematic quality-oriented approach for the analysis of laser beam ablation of this work similarly targets a quality-oriented flexibility for changing laser system environments.

This chapter 3 is situated in the first phase *plan* of the structure of this thesis (Figure 4). It introduces a research concept (strategy) and two approaches of the work as a unit before going into detail in the following chapters, which set the basis for the practical part of the thesis. To have a solid understanding of the challenge that represents laser beam ablation with picosecond lasers, it is important to expose the positive and negative aspects surrounding the process. For this case, a strengths, weaknesses, opportunities and threats analysis (SWOT analysis) has been developed in figure 30.

Mentioned in chapter 2.1.4, the advantages or strengths of picosecond lasers, on one hand, include a pulse length of approximately 10 ps, a pulse intensity reaching approximately 10^{15} W/cm^2 and allowing an almost cold ablation in metals and other materials. Moreover, the laser system is reliable, stable, economical (when compared to shorter pulse laser systems) and offers an ease of system component integration, like in the case of optical components, such as, a theta lens. Picosecond laser systems offer a high power, manageable pulse repetition rates, a range of wavelengths, burst-mode and the possibility of focusing the beam in the range of µm. On the other hand, weaknesses of such systems

J. P. Calderón Urbina, *Efficient material laser beam ablation with a picosecond laser*, Light Engineering für die Praxis, https://doi.org/10.1007/978-3-662-61886-8_3

are linked to the life cycle of traditional technologies, such as, grinding and chipping, among others, which present faster processing times, economic systems, higher processing rates and more experience. Another aspect that weakens the use of laser systems is the necessary investigation of parameters and settings to process a new or unknown material.

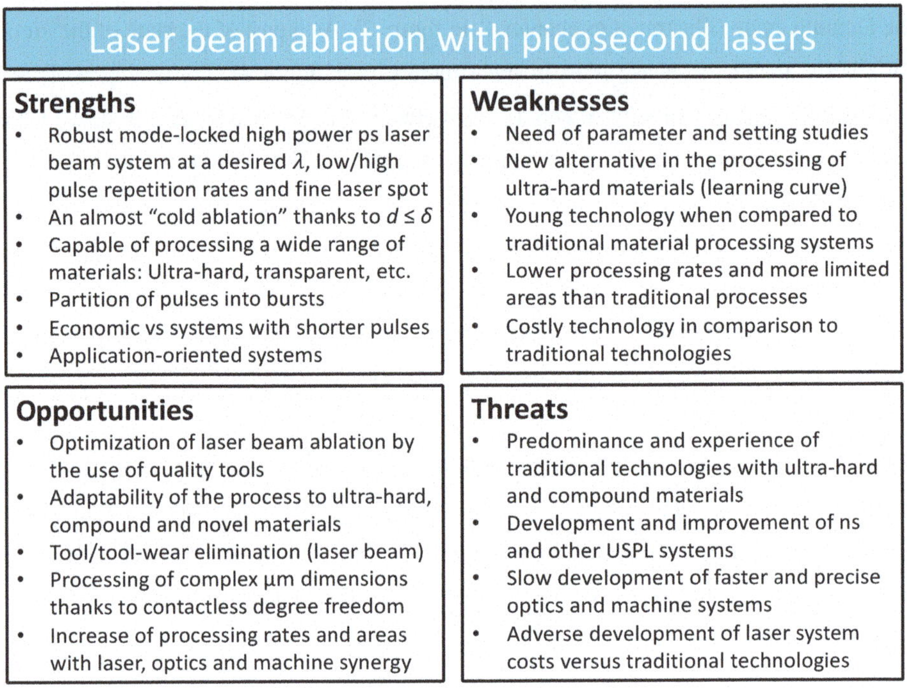

Figure 30: SWOT analysis

Among the threats to picosecond lasers are the development of other ultra-short pulse systems, such as fs and ns, and the predominance of traditional technologies. The advances in competing USPL technologies, such as fs and ns, are the generation of higher intensities and the improvement of precision for the first, and reliable cheaper systems for the latter. Likewise, traditional technologies could represent a minimum but important risk to the development of picosecond lasers because of their experience, results and costs. The development of fundamental elements, such as, optics and machine, accelerates or stops the positive development of lasers. The development of new materials and applications could impact negatively the application of picosecond lasers if it is not correctly confronted, making a resource demanding learning curve. Finally, instable financial and economic aspects can always impact raw materials and emerging technologies. An important aspect of the threats is that they also mean opportunities to ps laser. For example, new materials demand higher qualitative standards and some of them are difficult or costly to process by traditional technologies; furthermore, the miniaturization of applications areas, like the nano technological fields, required tools that are able to process in the micrometer range, with determined degrees of freedom, tool-free or with complex dimensions, making picosecond laser attractive to enter other applications areas, such as, medical and implantology, photovoltaics, and precision tool fabrication, among others. Moreover, the use of quality-oriented tools supports the identification of significant parameters and settings to adapt the laser beam to ultra-hard and novel materials. In short, there are opportunities to benefit from ps lasers, especially, when the trend projects faster and precise optical systems, efficient laser sources and cost reductions.

The core part of this work seeks an *efficient* material laser beam ablation, understood as, a productive laser beam processing of materials maximizing resources and minimizing waste and losses to achieve desired results or goals. This understanding must be reached by proposing a strategy that identifies relevant parameters in the picosecond laser system, obtains data and information of them, analytically understand the process, optimizes their interactions to laser beam ablation, reviews results and improves them (optimal results and continuous improvement). This strategy targets to save research time and cost by implementing quality tools; moreover, it represents a flexible approach, which can be adapted to different materials, laser systems and processes.

Figure 31 schematically illustrates a process chain for an efficient laser beam ablation with a picosecond laser. It describes the strategy mentioned in the last paragraph complemented by a continuous optimization of resources, e.g., materials (cemented WC, PCD, etc.), processing time, machine operation time, operator time, etc. In the lowest part of the figure, the main approaches of this work and the practical application of laser beam ablation are presented.

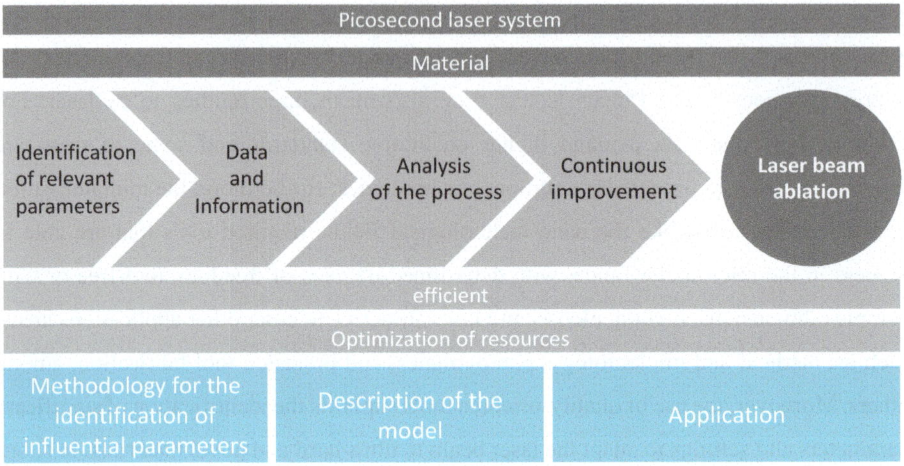

Figure 31: Process chain for an efficient laser beam ablation with a picosecond laser

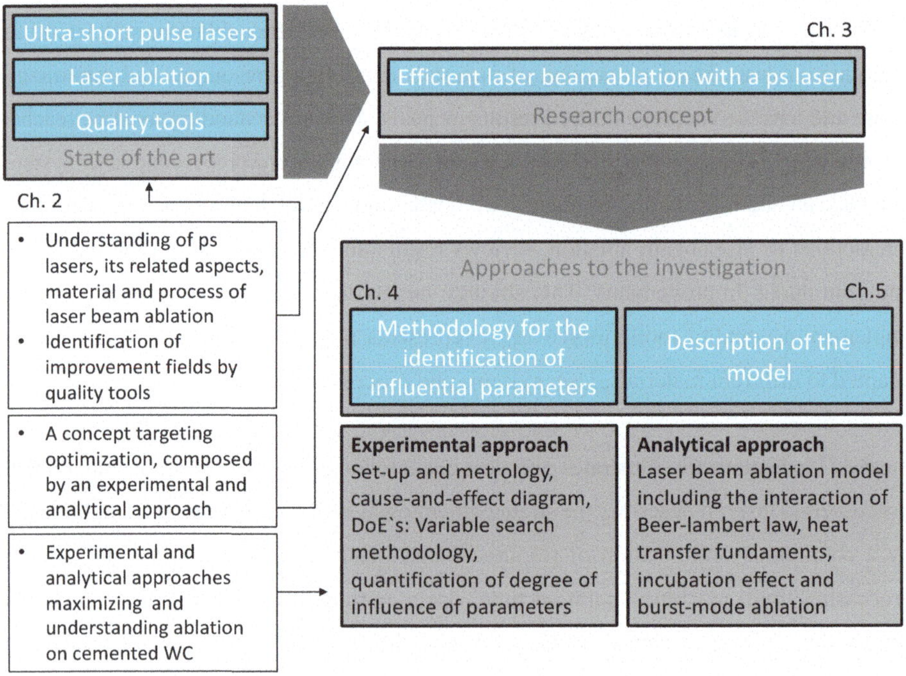

Figure 32: Methodical approach for the analysis of laser beam ablation

It can be observed the relationship of the methodology (identification of influential parameters) to the identification of relevant parameters, data and information, and the relationship of the description of the model to the analysis of the process. Thus, the approaches of this work contribute in the development of the process chain.

The research concept of this thesis conglomerates and joins the fundamental topics for an efficient laser beam ablation with an ultra-short laser, i.e., picosecond laser, to gain an understanding in both lasers and material ablation and optimize it with the use of quality tools, which seek for improvement fields. Figure 32 presents this in the block *state of the art*. Thus, a concept can be presented having two important components to target an *efficient* process and result. These two components are the *methodology for the identification of influential parameters* and the *description of a model*. The first component is an experimental approach looking for the most important and relevant parameter influencing the ablation process; the second component represents an analytical approach, which analyses from the theory the information collected from the experimental approach and targets the description of the ablation process by means of a model. The application of the cause-and-effect diagram, variable search methodology and quantification of the degree of influence of analyzed parameters on the ablation process of cemented WC with a picosecond laser are developed in chapter 4 as part of the experimental approach. The description of a model, as part of the analytical approach, takes into consideration not only the ablation process itself and heat transfer, but also the influence of the Beer-lambert law, the incubation effect and special processing strategies like single-pulse, multi-pulse and burst-mode laser beam ablation.

With this research concept, it is expected to define a strategy and action plan to set the basis for the operation, and thus, acquire enough information to define an optimal and efficient laser beam ablation strategy that continuously improves the application of picosecond lasers on cemented WC and other materials.

4 Methodology for the identification of influential parameters

This chapter presents the first approach of the research concept, a methodology for the identification of influential parameters of laser beam ablation on cemented WC. It serves as a guideline for the execution of the experiments further in this chapter, and thus, the determination of the influential parameters of laser beam ablation for further analysis.

Chapter 4 belongs to the first phase *plan* and the second phase *do* of the work and presents the methodology that is customized by quality tools and DoE's to identify the relevant factors of this investigation in laser beam ablation (see figure 4). The fundamentals of this chapter are focused on the definition of a strategy to identify the influential parameters (phase *plan*); the output of the methodology is based on the execution of the strategy (phase *do*). Based on the concepts of chapter 3, this part takes mainly advantage of the systematic approach of the cause-and-effect diagram and the variable search methodology. A factorial investigation of relevant parameters can only be done after obtaining the conclusions of a variable search methodology.

A systematic identification of influencing factors on the ablation process takes into consideration all possible factors related to the laser beam ablation system, paying special attention to those affecting directly the process, i.e., objectives and goals (for this work, the maximization of the ablation rate, mm^3/s). Thus, the understanding of the system is gained, and a consolidation of important factors is done. As described in subchapter 2.4.2, a cause-and-effect diagram supports the understanding of a system, and this understanding is supported in this work by interviews, collaboration and involvement in the operation of the system components. This knowledge leads to select a set of parameters which might have a relevant influence on the objective and that can be controlled. In this part, the application of the variable search methodology focuses on the determination of the degree of influence of parameters to select and discard the ones with the highest and the lowest or null influence, respectively. The choice of the parameters and the selection of the settings are based on a prior knowledge or an intuitive direction, the objective of the

J. P. Calderón Urbina, *Efficient material laser beam ablation with a picosecond laser*, Light Engineering für die Praxis, https://doi.org/10.1007/978-3-662-61886-8_4

methodology is to corroborate or discard these knowledge or direction in cases where not much information is available. One of the main advantages of this part is the reduction in experimentation, and consequently, time and costs. Finally, as schematically seen in figure 33, a set of main influencing factors is obtained to fine tune their impact on the objective. It helps to understand the interactions and contrasts between them, and to be able to optimize the process and its results.

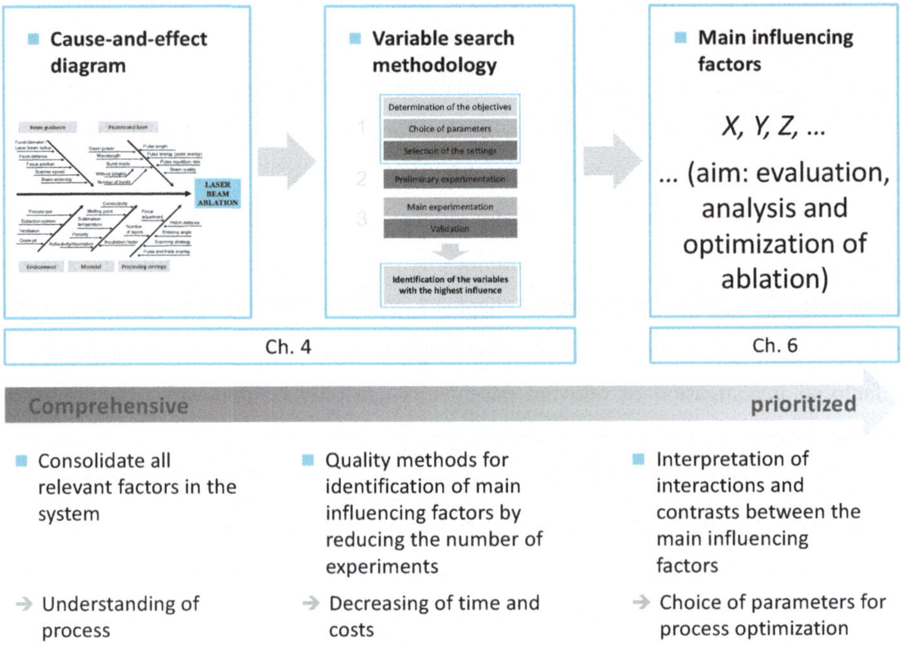

Figure 33: Systematic identification of influencing factors on the ablation process

The following lines describe in a detailed manner the laser beam system and the selected quality tools from a general (considering the complete laser beam system and its optics, laser, environment, material and strategy parameters) to a case-oriented perspective: Laser beam ablation of cemented tungsten carbide (WC) with a picosecond laser. The target of the methods of the systematic identification of influential parameters is the maximization of ablation rate.

4.1 Cause-and-effect diagram for picosecond laser beam ablation of WC

Chapter 2.4 introduced the cause-and-effect diagram for laser beam ablation (figure 26). The following lines describe the diagram by emphasizing the relevant fields for this investigation and selecting the parameters (causes for Ishikawa) that might represent an influence on the maximization of ablation rate on cemented WC (effect for Ishikawa).

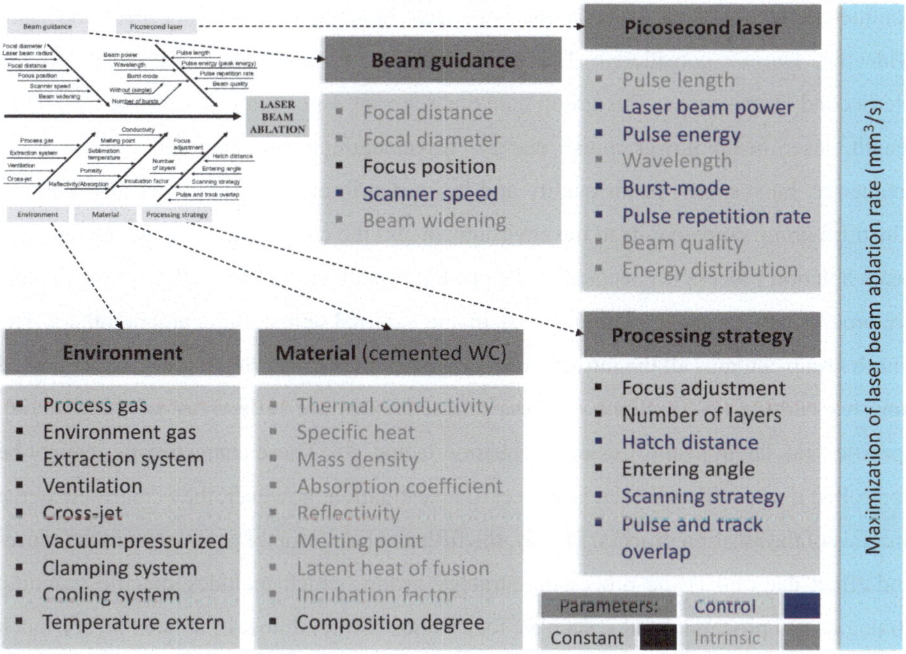

Figure 34: Relevant factors for picosecond laser beam ablation of WC from the cause-and-effect diagram

In a laser system, there are parameters that can be controllable and non-controllable. There are others that keep the system functioning just with their presence and others that can be discarded and do not affect certain objective. Thus, for this work the parameters were classified into control, constant and intrinsic. The control parameters are the selection of parameters under investigation that are set to desired different levels. The constant parameters are the selection of parameters for this work that are kept constant in a selected level. The last classification, intrinsic parameters, represents fundamental characteristics of the involved elements. From analysis of the cause-and-effect diagram for this work, five groups of parameters that might affect the objective of maximizing the picosecond laser beam ablation rate on cemented WC were defined (as depicted in figure 34). The first one corresponds to the beam guidance or beam delivery system, responsible of bringing the laser beam on the work piece from the laser source. To this matter, parameters, such as, focal distance, focal diameter, focus position, scanner speed and beam widening, can be considered. The second group of parameters corresponds to the picosecond laser source. This group contains relevant aspects for this thesis, e.g., pulse length, laser beam power, pulse energy, wavelength, pulse repetition rate or pulse frequency, burst-mode, beam quality and beam distribution (pulse power distribution). The third group corresponds to the environment and the circumstances in which the laser beam ablation process is executed in the machine chamber. Example of these parameters are process gas, cross-jet, clamping system and external temperature, among others. The fourth group contains all the intrinsic characteristics of the material, in this case cemented tungsten carbide (WC). Some of these characteristics are reflectivity of the material, specific heat, latent heat of fusion, incubation factor, etc. These characteristics cannot be controlled if a specific material is subject to investigation; however, they contribute to the analysis of the ablation process. Finally, the fifth and last group of parameters in the cause-and-effect diagram is the processing strategy group, which includes specific scanning strategies and processing runs, such as, focus adjustment, number of ablated layers, hatch distance, beam incidence angle, and pulse and track overlap of the subsequent pulses in a run.

Figure 34 presents in dark gray the parameters that are constant and in light gray the parameters that are intrinsic characteristics of the system elements, such as, material characteristics. Moreover, the figure remarks in blue the parameters that can be controlled (control parameters) and are considered relevant for a further investigation possibly having an impact in the objective. Those parameters are scanner speed v_s, laser beam power P_L, pulse energy E_0, burst-mode n, pulse repetition rate f, scanning strategy, and pulse and track overlap *PO*. This set of parameters is explained in the following subchapter, experimental set-up, making the concepts tangible.

In short, the presented cause-and-effect diagram for the picosecond laser beam ablation of WC offers a wide perspective of the system and targets the understanding of relevant factors that might influence the maximization of the ablation rate. Thus, these factors can consciously be neglected or included in the experimental approach and its further analysis. The determination and selection of the parameters is achieved in the following parts of this chapter.

4.2 Experimental set-up

The experimental set-up for the picosecond laser beam ablation of cemented WC is composed by parameters belonging to the five relevant fields illustrated in the cause-and-effect diagram. Table 4 presents a list of all the system fields and their parameters, and it classifies them accordingly to their role to this work in control, constant and intrinsic parameters. In practical terms, control parameters mean factors with the possibility of being directly manipulated; intrinsic parameters represent inherent characteristics of the factors or properties that cannot be manipulated. Constant parameters represent controllable parameters that can be set to a fixed desired level or selected under desired characteristics but are kept constant for the experimentation due to expected experimental conditions, lack of relevance for the investigation, or economical aspects.

In beam guidance, scanner speed v_s is selected as a control parameter thanks to its interaction with hatch distance Δ_y (processing strategy) and pulse repetition rate f (picosecond laser) to generate pulse and track overlap; a relevant aspect for the experiments and explained in 4.3. Focus position z_f corresponds to the position of the laser spot on or in the material and for this work is kept constant after investigating experimentally its low impact on ablation rate [8].

Table 4: Classification of system parameters

Parameters	Remarks	Control	Intrinsic	Constant
Beam guidance				
Focal distance f_d	Determined by scanner		•	
Focal diameter d_w	Determined by scanner		•	
Focus position z_f	Position: focus / material			•
Scanner speed v_s	Determined by scanner	•		
Beam widening φ	Determined by scanner		•	
Picosecond laser				
Pulse length τ_H	Laser source: 10 ps		•	
Laser beam power P_L	Average beam P of 50 W	•		
Pulse energy E_0	In relationship with P_L	•		
Wavelength λ	Infrared 1064 nm		•	
Mode (single/burst) n	Pulse groups of ns ranges	•		
Pulse repetition rate f	From 400 to 1000 kHz	•		
Beam quality M^2	Determined by source		•	
Energy distribution -	Determined by source		•	
Environment				
Process gas -	Machine standard			•
Environment gas -	Breathable			•
Extraction system -	Machine standard			•
Ventilation -	Machine standard			•
Cross-jet -	On in processing			•
Vacuum-pressurized -	Machine standard			•
Clamping system -	Adapted to WC/PCD			•
Cooling system -	Machine standard			•
Temperature extern T_r	From 20 to 25 °C			•
Material				
Thermal conductivity k	Thermal property		•	
Specific heat C_p	Thermal property		•	
Mass density ρ	Volumetric mass density		•	
Absorption coef. α	Optical property		•	
Reflectivity R	Optical property		•	
Melting point T_m	Thermal property		•	
Latent heat of fusion LH_f	Thermal property		•	
Incubation factor S	Accumulation effect		•	
Composition degree -	12% Co 88% WC			•
Processing strategy				
Focus adjustment z_{fa}	Executed by machine axis			•
Number of layers n_L	Executed by scanner			•
Hatch distance Δ_y	Determined by scanner	•		
Incidence angle γ	Workpiece vs laser beam			•
Scanning strategy $ScSt$	Scanning path	•		
Pulse/track overlap PO	Based on DIN 32540	•		

The rest of the parameters in beam guidance belong to the intrinsic characteristics of the optical system. The group of controllable parameters under picosecond laser represents the highest number in this research. Its importance for its understanding makes the controllable parameter a focus of the investigation. Laser beam power P_L directly related to pulse energy, is a parameter that influences the ablation process and requires high attention; single mode or burst-mode *n*, as describe in chapter 2, corresponds to a partition of a pulse in a form of irradiation of micro pulses or bursts separated in nanosecond range under a selected pulse repetition rate. Pulse repetition rate *f* or pulse frequency is a factor that determines the rhythm of the pulses and interacts, as mentioned before, with scanner speed and hatch distance to achieve pulse overlap and track overlap of the pulses according to DIN 32540. Pulse length τ_H, wavelength λ, beam quality M^2 and energy distribution are intrinsic characteristics of the laser source. The environment, which represents an industrial laboratory environment keeps all their parameter constant following standard machine operation. The characteristics of the material correspond to an important part of the analytical investigation of this work and they are all intrinsic properties of the selected composition of WC. Finally, under processing strategy, pulse *PO* and track overlap *TO* are control parameters that define how much energy is delivered to the material; hatch distance is similarly a control parameter, but it is set accordingly to a targeted *PO* and *TO*. Scanning strategy defines the way the laser beam is driven on the material and it can be controlled due to the number of different available options in the system; some of the options of laser beam trajectory are unidirectional, bidirectional and meandering. The rest of the parameters in the group of processing strategy are kept constant after preliminary experimentation and evaluation.

4.2.1 System equipment

The set-up of the system is constituted in practical terms in three main components: a laser system, a scanner system and a machine system. The first component corresponds to a picosecond laser beam source of the company LUMERA (now Coherent, Inc., Santa Clara, California, U.S.A.) [129], model Hyper Rapid 50 at a wavelength λ = 1064 nm, pulse length of $\tau_H = 10$ ps, a focal diameter on working area of $d_w = 28$ µm, and an average power beam $P_L = 50$ W at $f = 1000$ kHz, among other frequencies (table 5 complements the specifications).

Table 5: Laser beam source parameters

Wavelength $\lambda = 1064$ nm, focal diameter on working area $d_w = 28$ µm			
Pulse repetition rate f	Average beam power P_L	Pulse energy E_0	Laser fluence F
[kHz]	[W]	[µJ]	[kJ/m^2]
400	50	125	203
500	50	100	162.4
625	50	80	129.9
800	50	63	101.5
1000	50	50	81.2
TEM00 $M^2 < 1.5$		Pulse length $\tau_H = 10$ ps	

Measurements executed on the machine registered the following characteristics: output average beam power, focal diameter, beam quality and Rayleigh range (see table 6). In the first part of the table, it can be observed the characterization of the picosecond laser beam in terms of diameter and radius, quality and Rayleigh range.

The second part of the table presents the measurement of the output average laser beam power from two different positions, the first, measurement a, from a position close to the beam focus (+/- 10 mm). The measurement exactly on the focus was avoided because it represents a risk for the measurement device. The second position, measurement b, is done directly under the scanner.

Table 6: Local measurements of the picosecond laser system

Picosecond laser system at a wavelength λ = 1064 nm					
Selected average beam power P_L	Selected pulse repetition rate f	Laser beam radius ω_0	Focal diameter d_w	Beam quality factor M^2	Rayleigh range $z_{R, real}$
[%]	[kHz]	[µm]	[µm]	[-]	[mm]
100	400	13.9	27.8	1.17	0.49
100	500	13.7	27.4	1.18	0.47
100	625	13.1	26.2	1.15	0.44
100	800	13.5	27.1	1.21	0.45
100	1000	13.5	27	1.25	0.44
Focal length 163 mm, single pulse (without burst-mode)					

Selected average beam power P_L	Measured average beam power P_L	Selected average beam power P_L	Measured average beam power P_L	Considerations
[%]	[W]	[%]	[W]	
80	45.7	80	47.3	The average beam power measurement was executed (a) close to the focus (considering focal length) and (b) directly under the scanner in two different sessions, both at a f = 400 kHz and single pulse (without burst-mode).
85	47.2	85	49.4	
90	48.2	90	50.8	
95	48.6	95	51.7	
100	48.4	100	51.9	
Measurement a		Measurement b		

Figure 35 illustrates the characterization of the beam, i.e., focal diameter, beam quality and Rayleigh range, under different pulse repetition rates based on table 6. There is a relatively constant behavior of the results; however, there are minimal and inherent fluctuations due to measurement conditions and environment. An important characteristic of the beam for this work is focal diameter; its measurement fluctuates from d_w = 27 to 27.8 µm. For this work, the focal diameter is considered as d_w = 28 µm, including aspects, such as, measurement conditions, range of magnitude (µm) and minimization of computation. The average beam quality factor from the five measurements of table 6 is of 1.19 and the average Rayleigh range is of $z_{R,\,real} = 0.45$ mm.

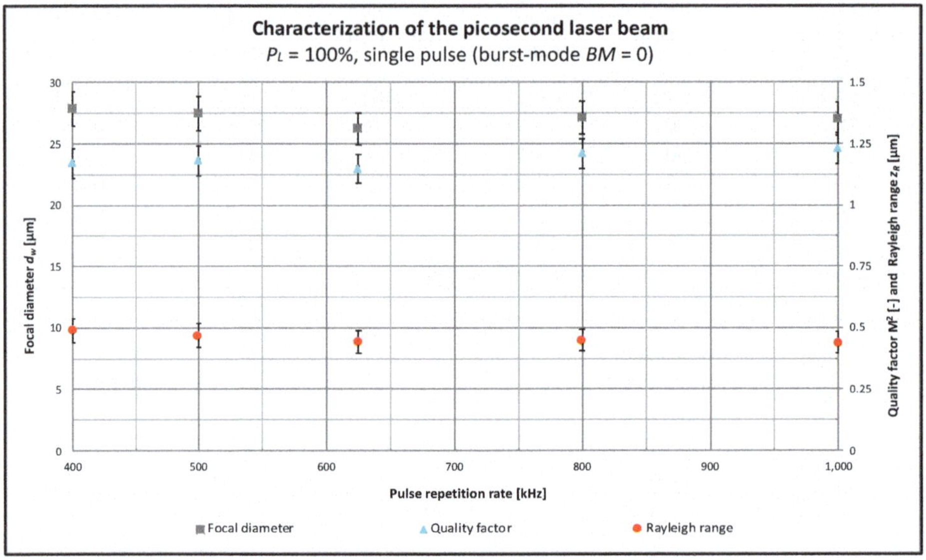

Figure 35: Characterization of the picosecond laser beam

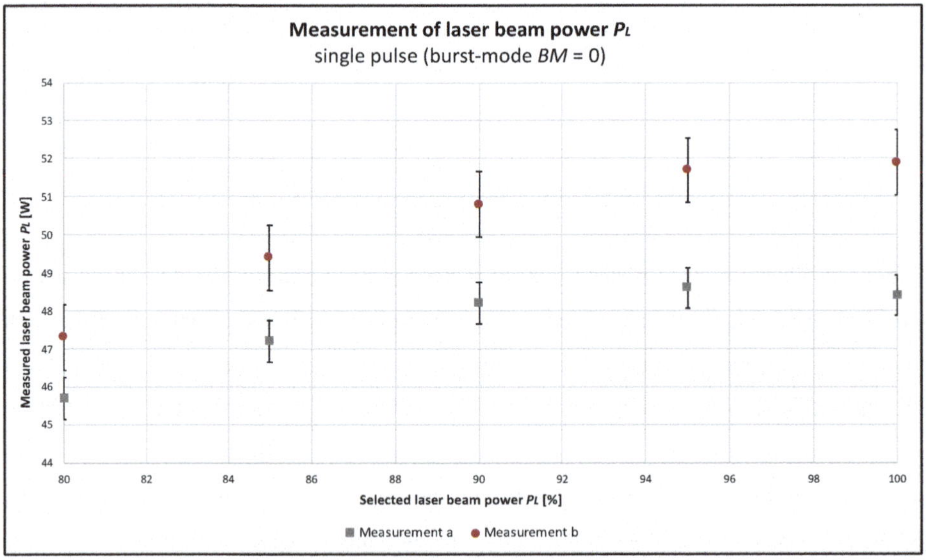

Figure 36: Measurement of laser beam power P_L

The measurement of the laser beam power is illustrated in figure 36. The measurement was executed from 80 to 100% in 5% increments and, as mentioned before, in two different positions, close to the focus and under the scanner, measurement a and measurement b, respectively. Taking into considerations the different measurements, the influence of the measurement conditions and environment, the full laser beam power of 100% is considered and adapted for this work as P_L = 50W. The characterization of the laser beam and the measurement of the average beam power were done using single pulse without a partition of it in bursts. The characterization used an average laser beam power of 100% in the five available pulse repetition rates (400, 500, 625, 800 and 1000 kHz).

The beam delivery is achieved by the implementation of a scanner system of the company ScanLab. Specifically, A hurryScanII is used and it offers a typical marking speed of 3.5 m/s and a positioning speed (scanner speed) of v_s = 15 m/s. It offers a 3-axes scanning system (X, Y, and Z). The third and last component, the machine system, contains and integrates the other two systems under a 5-axes movement that enables, e.g., programming, focus adjusting, angles, residual product extraction, etc.

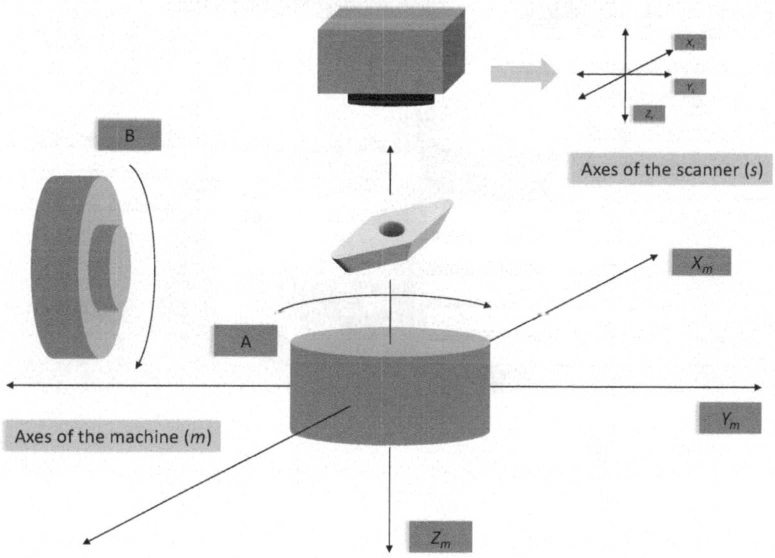

Figure 37: Machine axes system

The model Laser Line of the company EWAG integrates coordinately all the commands for laser processing with the laser beam source, the scanner system, and its machine axes, as seen in figure 37.

Figure 38 illustrates schematically the experimental set-up conditions: A picosecond laser beam source, a processing chamber, where the laser processing is done, composed of a scanner system and a work piece table with 5-axes, and lastly, a machine control that integrates all the processing preparations, execution and release. The image of the processing chamber shows schematically a laser beam processing of a cemented tungsten carbide tool accompanied of residual extraction, illumination, and 5-axes movement.

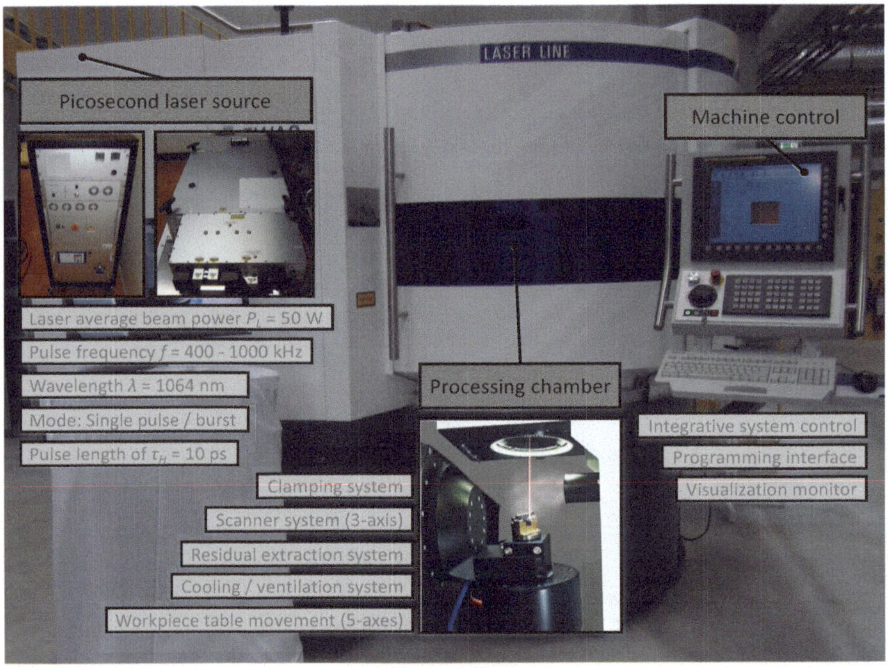

Figure 38: System equipment: laser beam source, scanner system and machine system

4.2.2 Material and sample preparation

The material under investigation in this work is cemented tungsten carbide WC chemically composed of Cobalt Co and tungsten carbide WC. Its composition is divided into 12% Co and 88% WC; it presents a grain size of 0.5 μm and it is treated as a composite material (cemented tungsten carbide), which is processed homogenously and not individually in single Co or WC. For the experiments, a square blank of standard cemented WC is used. The dimensions of the blank sample are 13 mm x 13 mm x 5 mm (length x width x height). There is an orifice in the middle of the blank, which corresponds to the clamping orifice used by a, e.g., standard milling CNC machine, with a radius of r = 2.5 mm. Figure 39 shows a sample of a standard cemented tungsten carbide blank. It can be appreciated in figure the surface and texture due to fabrication of the blank. The preparation of the sample for the experiments is simple. It requires an alcohol-based cleaning with a soft wipe; it is important to avoid the contact of extra substances to its surface. After the experiments, it is necessary to clean the sample of residues without influencing the surface. A further and deeper description of the intrinsic properties of WC is developed in chapter 5 for the analytical approach.

Figure 39: Sample of a standard cemented tungsten carbide WC blank

4.2.3 Experimental procedure

To execute the experiments of the variable search methodology, it is important to define an ablation experimental procedure of the specific ablation operation. This is a basic ablation form that is used for all the ablation runs and contains constant reliable characteristics. For this work a square cavity shape of 4 mm^2 on the blank was selected for ablation (area of 2 x 2 mm^2) to optimize the use of the material samples, maximize the number of cavities, avoid waste and allow confocal laser scanning measurements. For the analysis of the selected parameters of the methodology and based on previous experiments, preparatory experiments and investigations [8; 19; 88] number of layers of irradiation n_L and focus position z_f are kept constant. Figure 40 and 41 describe schematically an ablation cavity for this work; the first figure presents an actual picosecond laser beam ablated cavity, while the second figure illustrates a schema of the ablated volume.

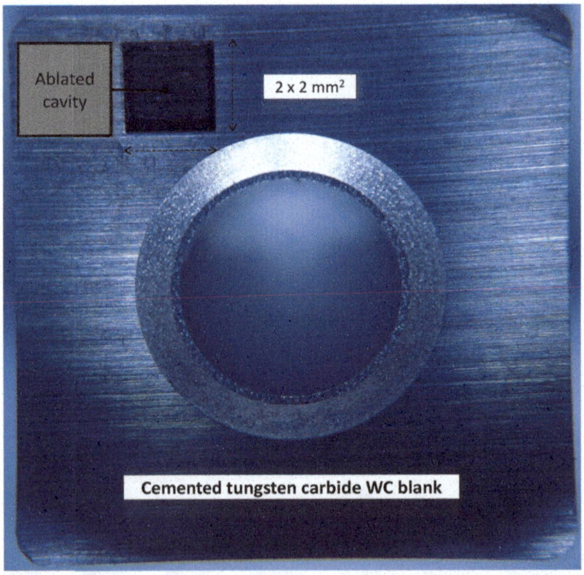

Figure 40: Picosecond laser beam ablated cavity on cemented WC

The measurement of the ablation depth Z_{abl} and the confirmation of the dimensions of the planned area contrasted to the actual ablated area, define the ablated volume, which represents an important fact to calculate the ablation rate with the processing time. Figure 41 exemplifies the notion of an ablated volume on a surface.

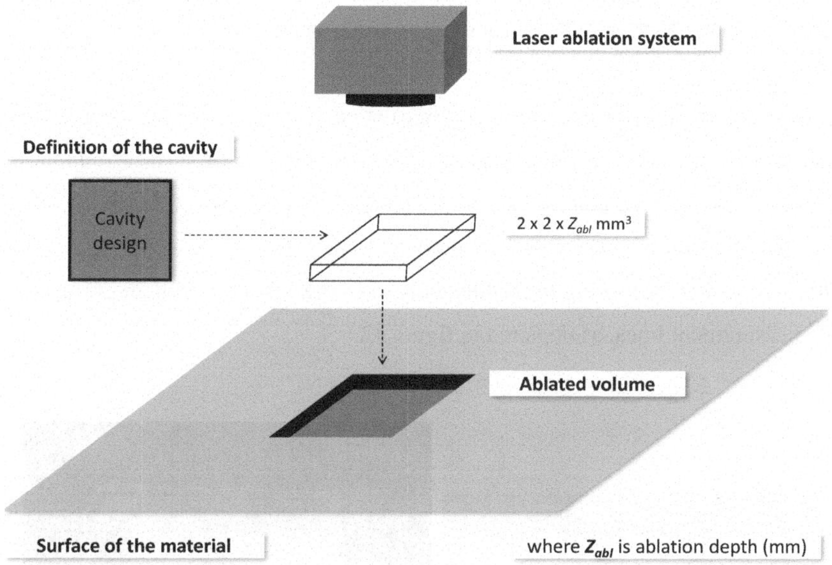

Figure 41: Picosecond laser beam ablated cavity on cemented WC (scheme)

The experimental set-up described in this part represents an important basis for the understanding and execution of the experiments of the variable search methodology. The following part, 4.3, deeply describes the variable search methodology (preparation and experimentation) adapted to the identification of the influential parameters on the picosecond laser beam ablation of cemented WC.

4.2.4 Metrology

The measurement of laser ablated cavities represents a challenge for conventional 2D microscopy. The measurement of depth and surface quality is possible by mechanical methods, but the use of optical techniques improves the ease and accuracy of the measurement. Therefore, an optical 3D laser scanning was used to measure the dimensions of the cavities in a sub-millimeter range. Specifically, the measurements for this investigation were done with a confocal laser scanning microscope VK 8710 Color 3D of the company Keyence (Keyence Corp., Osaka, Japan) [130].

The measurement procedure is executed after the cavity processing has been completed and the sample has been cleaned. Measurement and analysis of surface roughness and ablation depth is done with the software VK Analyzer. Roughness, if needed, is measured according to the surface roughness measurement (ISO 4287:1997), while the depth of the cavities is measured according to profile analysis of the software with an average profile of 400 measurement lines, as depicted in figure 42.

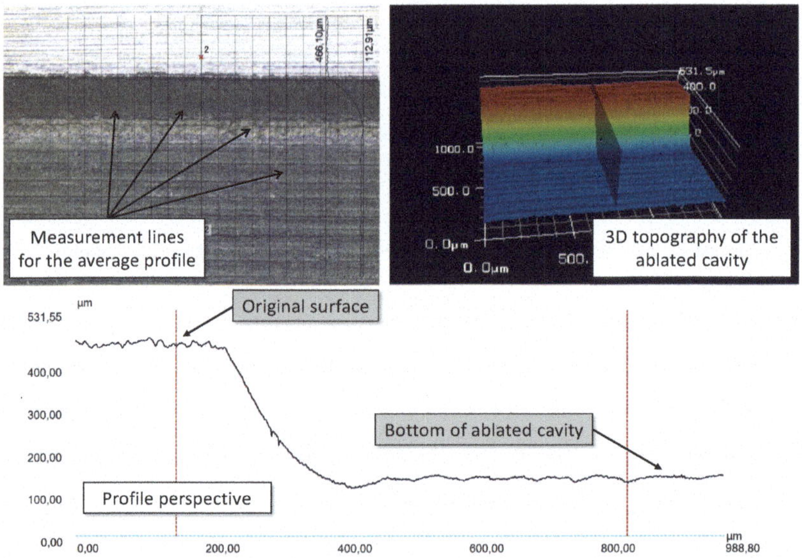

Figure 42: 3D confocal microscope measurement of an ablated cavity

Once the ablation depth is determined by the measurement, the ablation rate *AR* is calculated by equation 12:

$$AR = \frac{Z_{abl} \cdot L \cdot W}{t} \quad (12)$$

Where, AR, denotes ablation rate; Z_{abl}, ablation depth; L length of the ablation area; W width of the ablation area; and t, total processing time of the cavity sample.

4.3 Variable search methodology

The search of relevant process parameters in the picosecond laser ablation of cemented WC is initiated by the application of the variable search methodology or Shainin`s methodology. This methodology, explained in subchapter 2.4.3, quantifies the level of influence of laser and system parameters to ablation rate AR mm^3/s by contrasting several combinations of the parameters and their settings in the experimentation. The following lines develop the methodology and describe its parts by starting at the determination of the objective, following the choice of the parameters and their settings, the plan of the preliminary and main experimentation, the execution of the experiments, and ending with the presentation of the results. A conclusion of the results will be presented at the end of the chapter and a global analysis of the conclusions and the analytical approach will be shown in chapter 6.

The experimental set-up in 4.2 described the system equipment, material, sample preparation and experimental procedure that it is executed in the variable search methodology for each of the experiments. Likewise, the procedure of metrology for the measurement of the laser ablated cavities was presented.

4.3.1 Definition of the objective

The understanding of the behavior of picosecond laser beam ablation in this work will be transferred to the application of lasers of this kind to the processing of material in industrial environment. Thus, industrial environments demand an efficient processing of materials understood as a time-quality-cost relationship. For this reason, the objective of this methodology focuses in the maximization of the ablation rate of cemented WC and it is quantified by the measurement of the ablated cavity, as in figure 41, (mm^3) and its processing time (s). The objective of the variable search methodology for the identification of the influential parameters by Shainin in this work is:

The maximization of ablation rate AR, understood in this work as quantitative positive Qn+ and measured in volume per time (mm^3/s).

This searches for an increase of ablated volume, i.e., raise of ablation depth, of cemented WC through a cycle of time (processing time determined by the system parameters, e.g., scanner speed).

4.3.2 Selection of the parameters and settings

The selection of the parameter and settings must search the maximization of the ablation rate. This selection is done from the universe of parameters related to laser beam ablation in the system (grouped in the cause-and-effect diagram) and based on previous knowledge or experience, which is the case in this study with also other materials and objectives [19; 88; 131]. Three aspects must be observed: First, the selected parameters must be considered relevant to the objective in the system; second, their settings must influence the objective in a positive and negative response; and third, the parameters must have the possibility to be controlled. Basically, there must be two groups of settings, the first must set the parameters to achieve a maximization of the ablation rate of cemented WC (*Qn+*); the second group of parameters must have their settings intending a minimization or hindering of the ablation rate (*Qn-*). These two directions are needed to create a contrast in the response, and thus, identify the parameter or parameters with the greatest influence on the objective.

Before listing the selected controllable parameters of this investigation, it is necessary to mention important constant parameters and the integration of a set of parameters in pulse overlap *PO* and track overlap *TO*.

Constant parameters

Table 4, in subchapter 4.2, presented the constant parameters of the laser beam ablation system; such as, the environmental factors which represent standard conditions of a laser system (e.g., process gas, environment gas, ventilation, cross-jet on and room temperature) and the material with its properties and composition (subchapter 4.2.2). Likewise, there are other relevant constant parameters belonging to beam guidance and processing strategy; these parameters are listed and defined in the following table.

Table 7: Constant parameters in beam guidance and processing strategy

Parameter	Remarks	Value
Beam guidance		
Focus position z_f	Position: focus vs material	0 mm
Processing strategy		
Focus adjustment z_{fa}	Executed by machine axis	None
Number of layers n_L	Executed by scanner	50
Incidence angle γ	Laser beam to workpiece	90°

Previous studies [8; 19; 88] investigating ablation depth have shown slight favorable results when the beam focus is placed exactly on the material, $z_f = 0$ mm, and compared to beams above the material and in the material, e.g., $z_f = 1$ mm and $z_f = -1$ mm, respectively. Moreover, the influence of focus position does not increase significantly when its settings are varied. Therefore, $z_f = 0$ is kept constant for this study. Focus adjustment z_{fa} targets to keep the focus position at its original level, i.e., to adapt the focus position to the depth of the cavity after ablating layers, especially, when processing deep cavities ($Z_{abl} > 1$ or 2 mm). Nevertheless, focus adjustment is not included in the experimentation because the projected and preliminary ablation does not exceed a depth greater than the Rayleigh range z_R of the optical system (presented in table 6 and in the following lines) [24].

Equation 13 describes the Rayleigh range z_R in a Gaussian distribution of the beam from π, focus radius ω_0 and wavelength λ [18]; for this case, $\omega_0 = 14$ µm and $\lambda = 1064$ nm, the theoretical z_R equals 0.5787 mm. Equation 14 includes the quality of the beam using the beam quality factor M^2 [22; 132].

$$z_{R,\ Gaussian} = \frac{\pi\omega_0^2}{\lambda} \quad (13)$$

$$z_{R,\ real} = \frac{\pi\omega_0^2}{M^2\lambda} \quad (14)$$

Table 6 presented a list of five measurements of M^2 of the laser beam having an average value of $M^2 = 1.19$. Therefore, the Rayleigh range including the beam quality factor is of $z_{R,\ real} = 0.48$ mm. A comparison of the calculated Rayleigh ranges and the measured ranges (in average) in Table 6, $z_{R,\ real} = 0.45$ mm, concludes that an ablation depth less than 0.9176 mm does not require a focus adjustment, $z_{fa} = 0$.

The number of layers n_L, also understood as the number of runs that the scanner irradiates the target accomplishing a single selected processing strategy area, was kept constant in 50. Previous studies [8; 19] have shown that 50 layers contribute to present significant ablation results for metrology and analysis even when the parameters and their settings are not targeting high ablation. Thus, the irradiation number is the same for several sets of experiments. The last constant parameter corresponds to the incidence angle of the laser beam γ; it describes the position of the laser beam in axis z to the plane of the working area plane x-y, where the surface of the sample is positioned (see figure 37). It is intended to minimize a deformation of the focus in an elliptical form due to a deviation from 90°; therefore, the incidence angle is set to $\gamma = 90°$ to keep it perpendicular to the surface of the work piece.

Integration of pulse overlap and track overlap

Previous studies [88] have executed analysis of individual laser and scanner parameters separately, e.g., choice of the parameters in the variable search methodology. However, it has been observed a logical interaction of pulse repetition rate f, scanner speed v_s and hatch distance Δ_y in the manner they together deliver dynamically the laser beam power, i.e., energy, to the material. This interaction and how energy is delivered on the material present different levels of laser beam processing (laser beam ablation) of the material and this remarks its importance. Thus, the relationship of f, v_s and Δ_y can be expressed as one factor that describes total or partial overlap of laser pulses [15]. There are two types of overlap; the first is named pulse overlap *PO* and it corresponds to the degree in which a pulse is overlapped by a consecutive pulse following the same trajectory on the scanner speed direction. The second, track overlap *TO*, corresponds to the overlap of track of pulses following a trajectory and being perpendicular to pulse overlap. The degree of overlap in % is understood as the degree of coverage of the pulse, being 100% a complete coverage of a pulse by its subsequent pulse and 0% an individual pulse without any coverage. The following figures illustrate pulse overlap and track overlap.

Figure 43 presents a description of two pulses in movement (dynamically), which is also the fundament to describe pulse overlap. The pulses are ruled by scanner speed, pulse repetition rate (i.e., pulse period τ_P), focal diameter and pulse length. This relationship causes pulses to elongate on the workpiece as described by $v_s \cdot \tau_H$. Likewise, figure 43 illustrates pulse overlap *PO* exemplifying three consecutive pulses in the *X* direction.

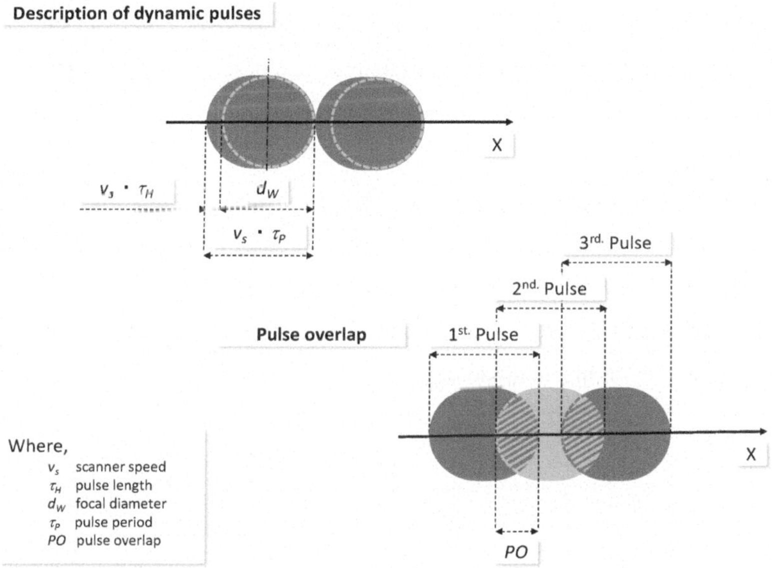

Figure 43: Pulse overlap [15]

Equation 15 determines the degree of *PO* based on scanner speed, pulse period, focal diameter and pulse length [15]:

$$PO = \left[1 - \frac{v_s \cdot \tau_p}{d_w + v_s \cdot \tau_H}\right] \tag{15}$$

The degree of pulse overlap can be determined from $PO = 0$, absence of overlap by separate pulses, to 100% overlap, $PO = 1$, when $v_s = 0$ m/s, and subsequent pulses cover 100% a preceding pulse. Track overlap *TO* is described by formula 16; it presents the relationship of hatch distance Δ_y and focal diameter d_w in the *Y* axis.

$$TO = \left[1 - \frac{\Delta_y}{d_w}\right] \tag{16}$$

TO represents the displacement in the *Y* axis of the track based on the hatch distance and the size of the focal diameter. A $TO = 1$ (100%) describes a track that did not displace from its preceding track and targets 100% this track; a $TO = 0$ corresponds to a track that does not overlap in any degree its preceding track, as in $\Delta_y \geq d_w$. Figure 44 illustrates the concept of track overlap by two track overlapping, and, describes the possibility of pulse and track overlap by overlapping preceding pulses and tracks.

The concepts of pulse overlap and track overlap are considered in this work to investigate their degree of influence in the variable search methodology and they are set as one factor with same magnitude $PO = TO$ for the analysis and experimentation. The component parameters of *PO* and *TO*, e.g., scanner speed, rate, pulse repetition rate, etc., are not included individually in the methodology.

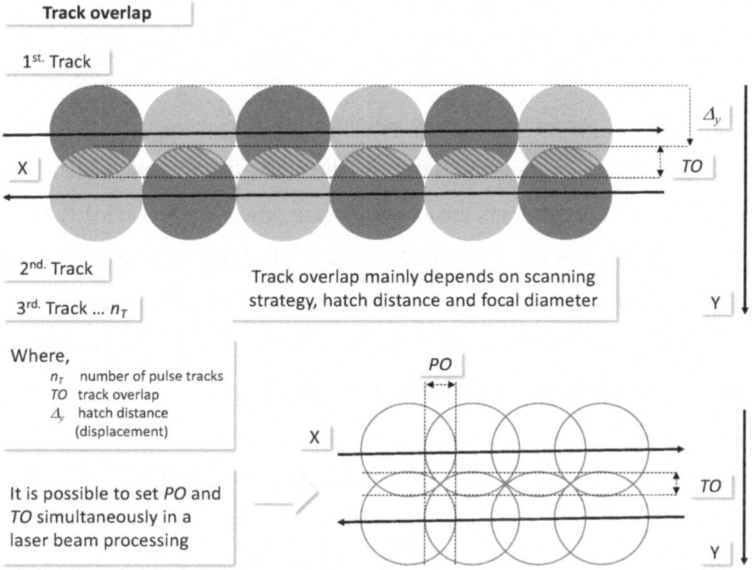

Figure 44: Track overlap

Choice of the controllable parameters and settings

The objective of the variable search methodology is to identify the parameters with the highest influence on ablation rate *AR Qn+* [mm³/s]. Before that, it is needed a selection or choice of the parameters and their settings that might affect the result and belong to the system under analysis. To achieve this, it is necessary to collect previous experience, information and knowledge, or simply select promising factors. For this study, a group of parameters was selected from the system projected in the cause-and effect- diagram thanks to their controllability and importance shown in previous studies with WC and other materials [8; 19; 88; 131]. The number of parameters was reduced to a minimum by a careful selection of them, and the determination of constant parameters. Their settings were chosen targeting a contrast in the objective to assure identification and gain information.

The selected parameters for the identification of their influence on ablation rate *AR* are four: Laser beam power P_L (factor A), burst-mode *n* (factor B), pulse overlap and track overlap *PO/TO* (factor C) and scanning strategy *ScSt* (factor D). The following lines complement the concept of the parameters.

Factor A, laser beam power P_L, energy per unit of time measured in watts [W] and belonging to the group picosecond laser in the classification of system parameters (experimental set-up). It represents the average beam power of the LUMERA picosecond laser with a maximum average power of P_L = 50 W. The power is set in percentage [%] in the machine and in watts in the methodology and analysis. The choice of settings for factor A presents a P_L = 25 W for *Qn-* and P_L = 50 W for *Qn+*, expecting to show a considerable contrast from low to high ablation, respectively.

Factor B, burst-mode *n*, explained in subchapter 2.5.1, corresponds to the capability of the laser to partition a pulse into a group of micro-pulses or bursts of nanosecond separations, following the pulse period dictated by the pulse repetition rate [88]. Burst-mode belongs, as P_L, to picosecond laser group in the classification of system parameters and it is dimensionless. The partition of pulses can be set from a quantity of 2 to a maximum of 10 micro-pulses, being 1 an absence of burst-mode, i.e., the application of single pulse without a partition. The use of factor B in the methodology intends to observe the influence of its application in the ablation process; for this reason, the choice of the settings includes a burst-mode of $n = 5$, and its absence, $n = 1$, i.e., single pulse. It is expected to achieve a higher rate of ablation by single pulse *Qn+* than by burst-mode *Qn-*.

Factor C, pulse overlap and track overlap *PO/TO* (for this study *PO* = *TO*), corresponds to a cluster of parameters which belong in the classification of system parameters to the beam guidance (focal diameter d_w, scanner speed v_s), picosecond laser (pulse repetition rate *f*) and processing strategy groups (hatch distance Δ_y). *PO/TO* describes, as mentioned before, the degree of overlap of consecutive pulses and track of pulses, and consequently, how pulse energy is delivered on the material. *PO/TO* is dimensionless [%] and has a range of *PO/TO* = 0 (no overlap of pulses) to 100% (complete overlap of pulse, only at v_s = 0 mm/s). For this investigation, the settings of factor C are *PO/TO* = 50% for *Qn-* and *PO/TO* = 90% for *Qn+*, respectively. The calculation of both overlap levels is done after equations 15 and 16 and it is presented in the following table.

Table 8: Values for the calculation of pulse overlap and track overlap *PO/TO*

Values for pulse overlap and track overlap *PO/TO*			
Desired *PO/TO*	Pulse repetition rate f	Scanner speed v_s	Hatch distance Δ_y
[%]	[kHz]	[mm/s]	[mm]
50	800	11200	0.01399
90	800	2240	0.00279
Focal diameter d_w = 28 µm		Pulse length τ_H = 10 ps	

Factor D, scanning processing strategy *ScSt*, a dimensionless factor, is a preconfigured irradiation trajectory of the laser spot that the laser scanner controls. It is found in the group of processing strategy in the classification of system parameters. Scanning processing strategy presents a relationship to hatch distance, i.e., track overlap, which defines the separation between the trajectories (pulse tracks) in an irradiation area. There are four basic scanning processing strategies: unidirectional, meandering pattern, zig-zag and wobble. Unidirectional strategy irradiates a track of pulses in one direction and jumping back without irradiation to the beginning of a new track to follow the same direction. Meandering strategy irradiates a meandering pattern which can be continuous, continuous irradiation, or with jumps, an interrupted irradiation between tracks, and also known as bidirectional strategy. Tracks irradiating diagonally from a point to another and similarly in the opposite direction behave in a zig-zag pattern. Zig-zag strategy maintains the laser continuously irradiating without jumps. The last basic or fundamental strategy is the wobble pattern; this strategy fills a form in a wobble or spiral trajectory, which follows the shape of the form. The laser is kept on and stops in the center of the form. A graphical summary of the fundamental scanning processing strategies of the system is presented in figure 45. There, it can be also observed two alternatives strategies based on the fundamental. One of them is a unidirectional strategy irradiating in a right to left direction, and the other, bidirectional strategy, is a meandering strategy with jumps between its lines. Previous studies indicated that unidirectional, meandering and bidirectional presented stable and favorable results in terms of quantity and quality (ablation rate *AR* and surface quality R_a, respectively), especially the last two [19; 131]; therefore, the choice of the settings of factor D are: *ScSt* = bidirectional for *Qn*- and *ScSt* = meandering for *Qn*+.

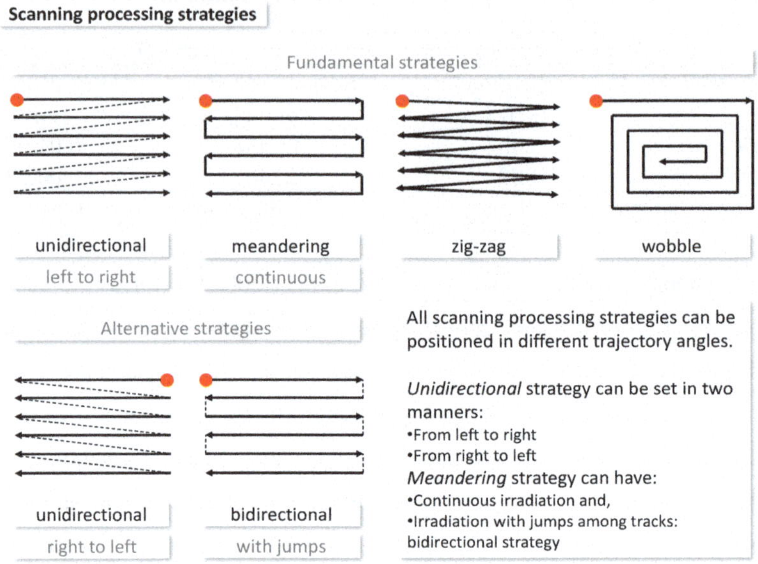

Figure 45: Scanning processing strategies

To conclude the selection of the parameter and settings, a group of controllable parameters has been chosen, and two series of settings has been defined; the first series targets maximization of the ablation rate of cemented WC (*Qn*+), and the second seeks a hindering of the ablation rate (*Qn*-). Table 9 presents the choice of the controllable parameter and their settings for the variable search methodology of this work.

Table 9: Choice of the controllable parameters and settings

Factors	Settings	
	Worst *Qn*-	Best *Qn*+
A Laser beam power P_L [W]	25	50
B Burst-mode n [-]	5	1*
C Pulse overlap and track overlap *PO/TO*♦ [%]	50	90
D Scanning processing strategy *ScSt* [-]	bidirectional	meandering

* Single pulse, $n = 1$

♦According to DIN 32540. Described in table 8 "Values for the calculation of pulse overlap and track overlap *PO/TO*", with the corresponding pulse repetition rate f, scanner speed v_s, and hatch distance Δ_y.

These parameters and groups of settings serve as basis of the variable search methodology in its following phases, as described in subchapter 2.4.

4.3.3 Preliminary experimentation and main experimentation

As mentioned in subchapter 2.4.3, the variable search methodology has two important parts after the parameters and setting were selected; the first is the preliminary experimentation, and the second part is the main experimentation. The preliminary experimentation represents an approach to obtain reference values for the main experimentation and its individual factor runs. The reference values help to recognize significant results of the parameter and setting combinations in a positive or negative direction of the objective, i.e., they are the highest and lowest response limits of the system. The main experimentation examines each parameter against the rest of them individually in both setting levels, quantitative positive *Qn+* and quantitative negative *Qn-*, setting first a parameter in a positive level against the rest of the others in a negative level, and consequently, setting this parameter in a negative level against the rest in a positive level, respectively. Finally, it quantifies its level of influence from the respective positive or negative reference value *V*. The following lines develop this knowledge in the laser beam ablation of cemented WC.

Preliminary experimentation

The parameters and settings from table 9 are prepared in a positive group (best *Qn+)* and negative group (worst *Qn-*) for experimentation with one replication each (first and second preliminary experiment). The negative group worst *Qn-* intends to present a discrete ablation rate, and the positive group best *Qn+* targets the maximization of it. The *AR* average results of the preliminary experimentation are identified as reference value V_+ and reference value V_-, respectively. Table 10 presents the results of the preliminary experiments and their average, which will be set as the reference value for both settings, respectively. The experiments gave a $V_+ = 0.0226$ mm³/s and a $V_- = 0.0078$ mm³/s.

Table 10: Preliminary experimentation results on laser beam ablation of WC (*Qn*)

Experiments	Results on *AR* [mm³/s]	
	Worst *Qn-*	Best *Qn+*
1st. Preliminary experiment	0.0084	0.0224
2nd. Preliminary experiment	0.0072	0.0228
Reference value *V*	0.0078 (V_-)	0.0226 (V_+)

Once the preliminary experiments are executed, it is important to do the D_s / d_s statistical test (based on the classical *F* table) of the variable search methodology to assure the presence of at least one dominant parameter in the methodology [13; 127]. The elements d_s and D_s correspond to the variation among repeat runs of experiments and the overall influence of all parameters changing their setting in the same period, respectively [13]. The test verifies the following relationship (cp. Subchapter 2.4.3):

$$\frac{D_s}{d_s} \geq \frac{5}{1} \tag{17}$$

In the case the D_s / d_s test fails, there are several reasons that might affect this result [13]. For example, the group of parameters was not adequately selected and it does not include a dominant factors or factors, the settings do not represent a high level of contrast or were not selected in the right positive/negative direction to the objective, or there are exclusive relationships of parameters that cause interactions between them, and thus, an effect manifestation when a parameter is set only in one direction.

The D_s / d_s test for the case of this work is executed in the following lines based on the results of table 10 and the equations 10 and 11 from subchapter 2.4.3.

Where for this case, B_1 and B_2 are the first and second *Best Qn+* preliminary experiment result, respectively; and W_1 and W_2 are the first and second *Worst Qn-* preliminary experiment result, respectively. The values of B_1, B_2, W_1 and W_2 are expressed in mm^3/s.

$$D_s = \frac{0.0224 + 0.0228}{2} mm^3/s - \frac{0.0084 + 0.0072}{2} mm^3/s; \tag{18}$$

$$D_s = 0.0148\, mm^3/s$$

$$d_s = \frac{0.0224 - 0.0228}{2} mm^3/s + \frac{0.0084 - 0.0072}{2} mm^3/s; \qquad (19)$$

$$d_s = 0.0004\, mm^3/s$$

$$\frac{D_s}{d_s} \geq \frac{5}{1}\,;\ \frac{0.0148}{0.0004} \geq \frac{5}{1}\,; \qquad (20)$$

$$37 \geq 5$$

The $D_s\,/\,d_s$ test criteria is fulfilled by $37 \geq 5$; therefore, the group of parameters and settings were correctly selected, and it is assured the presence of at least one significant parameter in the system (in Shainin´s nomenclature, a RED X factor).

Main experimentation

The examination of the parameters against the rest of them to identify a contrast in the results of the ablation rate is executed in the main experimentation of the variable search methodology (cp. Subchapter 2.4.3). Each parameter is tested independently in two parts: First, the parameter A_-, laser beam power P_L, is set at its negative level against the rest of the parameters R_+ set at their positive level. Second, the same parameter is set at its positive level A_+ against the rest of the parameters at their negative level R_-. A setting contrast in the parameters targets a manifestation of influential factors. After completing the experimentation with all the parameters at both settings, the results are registered and the variation to the corresponding reference values is quantified. Finally, the most significant parameters are identified after analyzing the degree of contrast. The parameters with the highest influence on the objective (response), "Red X" and "Pink X", are defined.

The test of the parameters is listed in table 11. It presents the combination of each of the parameters against the rest of them in both positive and negative settings, e.g., A_-R_+, A_+R_-, B_-R_+, B_+R_-, C_-R_+, etc.

Table 11: Main experimentation

Parameters	Combination	V_+ [mm³/s]	V_- [mm³/s]
A Laser beam power P_L [W]	$A_- R_+$	0.0226	-
	$A_+ R_-$	-	0.0078
B Burst-mode *n* [-]	$B_- R_+$	0.0226	-
	$B_+ R_-$	-	0.0078
C Pulse overlap and track overlap *PO/TO* [%]	$C_- R_+$	0.0226	-
	$C_+ R_-$	-	0.0078
D Scanning processing strategy *ScSt* [-]	$D_- R_+$	0.0226	-
	$D_+ R_-$	-	0.0078

Where, *R* represents the rest of the parameters, and V_+ and V_- the reference values for the best *Qn*+ and worst *Qn*- setting combinations, quantitative positive and quantitative negative, respectively.

Next to the parameter combinations can be found the reference values *V* (from table 10); each positive or negative reference value must conform to the positive or negative combination of rest of the parameters *R*, respectively. This is based on the variable search methodology´s fundament (after Shainin) of inducing a variation on the result when only one parameter is set in the opposite level than the rest. The result of each combination is compared to its respective reference values to quantify its level of influence. Results that are alike to their corresponding reference value might denote none or little influence of the factor under investigation in the target; on the other hand, a result presenting a greater difference would hint a positive or negative effect on the target depending on its influence direction. This is done in the execution of the experiment and results (subchapter 4.3.4).

4.3.4 Execution of the experiment and results

The runs of the experiments were executed on a cemented WC plate (see figure 46) and under controlled aspects described in the experimental set-up in subchapter 4.2 of this work. Measurements of the results are presented in the following table 12 for each of the combinations of parameters. The measurement of the cavity's depth is registered to calculate its volume by adding the cavity area (square cavity 4 mm²).

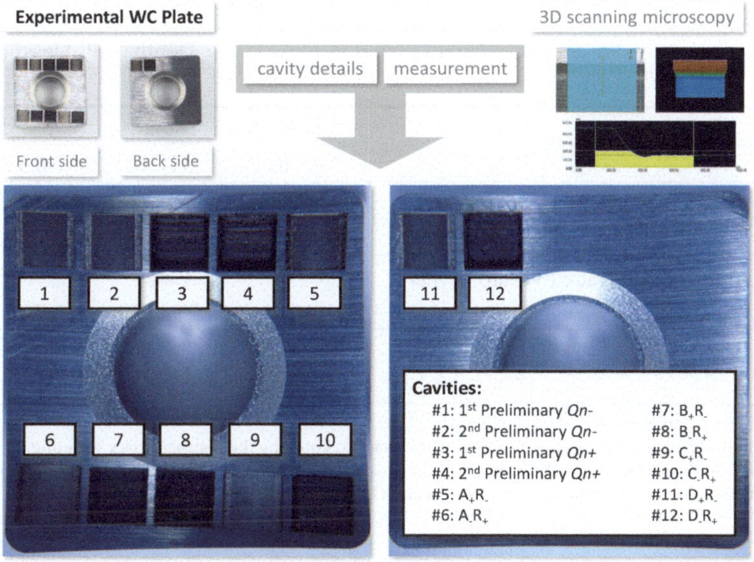

Figure 46: Results of the variable search methodology on a WC plate

The processing time corresponds to the time the laser beam ablates completely the cavity on the material from laser on to laser off. These two concepts, depth of the cavity and processing time, define the calculation of ablation rate for this methodology. Qualitative aspects of the cavities can be also noticed in figure 46. There, it is possible to have an initial observation of the influence of the parameter combinations of the methodology on the surface of the cavities. The quantitative aspects of the cavities are obtained by the measurement of the depth and the registration of the processing time, as presented in the following table.

Table 12: Measurements of cavity depth and processing time

Parameters	Combination	Depth [μm]	Processing time [s]
A Laser beam power P_L [W]	$A_- R_+$	186.28	50.352
	$A_+ R_-$	31.26	8.712
B Burst-mode *n* [-]	$B_- R_+$	418.52	50.273
	$B_+ R_-$	9.49	8.764
C Pulse overlap and track overlap *PO/TO* [%]	$C_- R_+$	12.51	8.653
	$C_+ R_-$	321.51	74.32
D Scanning processing strategy *ScSt* [-]	$D_- R_+$	266.5	74.296
	$D_+ R_-$	18.27	8.822

Where, *R* represents the rest of the parameters.
The depth of the cavity corresponds to the measurement executed by the confocal laser scanning microscope on the surface of the WC plate and the ground of the cavity. The processing time is registered by the machine from the beginning of the processing (laser on) until the end (laser off).

Table 13 presents the results of the experimentation for each of the combinations and expressed in terms of ablation rate [mm³/s]. Likewise, it is presented the absolute variation of each result when compared to its corresponding reference value and the average of the absolute variations for each parameter.

Table 13: Results of the main experimentation

Parameters	Comb.	V_+ [mm³/s]	V_- [mm³/s]	Results [mm³/s]	Var. %	Average %	Conclusion
A, Laser beam power P_L [W]	$A_- R_+$	0.0226	-	0.0148	35	59	*2. RED X*
	$A_+ R_-$	-	0.0078	0.0144	84		
B, Burst-mode *n* [-]	$B_- R_+$	0.0226	-	0.0333	47	46	*3. PINK X*
	$B_+ R_-$	-	0.0078	0.0043	45		
C, Overlap *PO/TO* [%]	$C_- R_+$	0.0226	-	0.0058	74	98	*1. RED X*
	$C_+ R_-$	-	0.0078	0.0173	121		
D, Strategy *ScSt* [-]	$D_- R_+$	0.0226	-	0.0143	37	21	*less significant*
	$D_+ R_-$	-	0.0078	0.0083	6		

Where, *R* represents the rest of the parameters, and V_+ and V_- the reference values for the best *Qn+* and worst *Qn-* setting combinations, quantitative positive and quantitative negative, respectively.
The variation is calculated based on the corresponding reference values. The variation (var.) of the results and its average are expressed in percentage [%].

The most significant variation of the results was presented with parameter C, pulse overlap and track overlap *PO/TO*, where its absence in a positive setting (*PO/TO* = 90%) decreased the result in approximately 74%; on the other hand, the application of a positive setting improved in approximately 121% the negative reference value. A second important parameter is the laser beam power (parameter A); a decrement of approximately 35% is expected, when laser beam power is set to 25W (negative setting) and an increment of ablation rate of approximately 84% is observed when laser beam power is set to 50W (positive setting). Parameter A and C are considered Red X parameters in the variable search methodology thanks to their level of influence in the ablation rate results. An interesting behavior is presented by parameter B, burst-mode. It shows an inverse interaction when it is set to an opposite value to the reference values, e.g., its negative setting B_- improves the result of ablation rate (positive reference value) in approximately 47%. Presenting a similar behavior but in the opposite direction, its positive setting B_+ decreases in approximately 45% the value of the negative reference value. This influence can be described as a Pink X parameter in the variable search methodology. Finally, parameter D, scanning processing strategy, presents less significant results when compared to other parameters in the methodology and only the absence of a meandering scanning strategy (like in the combination D_-R_+) shows a negative effect in the result. This might suggest an interaction of the meandering scanning strategy with the positive setting of some of the parameters.

4.5 Conclusions

A strategy to identify influential parameters was defined, *plan*, and its execution, *do*, was applied. The identification of influential parameters began from the identification and definition of several system parameters that were organized in the cause-and-effect diagram. There, it could be observed how many parameters can be involved in the laser beam ablation system and can have an effect in the output. Therefore, the determination of relevant parameters, such as, control, constant and intrinsic, was an essential milestone in the research. Based on that selection (control, constant and intrinsic parameters), an experimental set-up was defined fulfilling the conditions of a goal-oriented experimental investigation, i.e., maximization of the ablation rate. Thus, a variable search methodology for laser beam ablation on cemented WC was executed. Each of the parameters under investigation was tested against the rest of the parameters to identify its contrast in the result. Pulse overlap and track overlap *PO/TO* and laser beam power P_L (parameter C and A, respectively) were defined as the *Red X* of the system, i.e., the parameters with the highest influence on the ablation rate of cemented WC, presenting a level of influence (average variation to the corresponding reference value) of 98% and 59%, respectively.

Figure 47 shows the conclusions of the methodology and illustrates the behavior of the parameters when compared to their corresponding reference value.

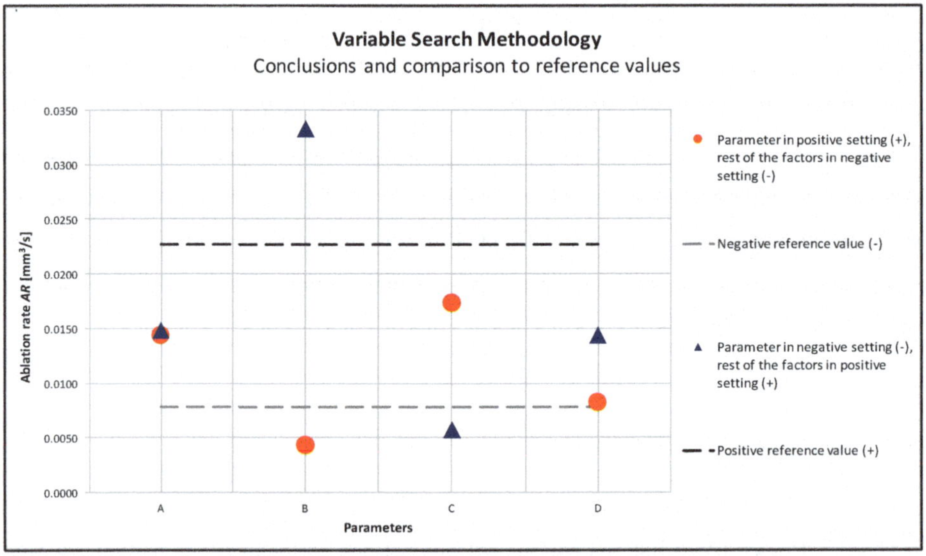

Figure 47: Conclusions of the variable search methodology

The *Red X* parameters, complemented by the *Pink X* (parameter B, burst-mode), represent the main elements to achieve an optimal ablation rate in the picosecond laser system. The interaction of influential parameters and component parameters, such as, pulse repetition rate f, scanner speed v_s and hatch distance Δ_y, in pulse overlap and track overlap *PO/TO*, defines the performance of the ablation. For this reason, the following approach in this work analyzes the interaction of these parameters with the behavior of the material and considers relevant laser beam conditions (e.g., incubation effect and Beer-Lambert law) to describe a model. In short, the following chapter describes and develops an analytical approach for picosecond laser beam ablation on cemented WC.

5 Description of the model

The fifth chapter of this work belongs to the second approach to this investigation after the experimental. It is situated in the first phase *plan* and the second phase *do* of the quality-oriented structure of the thesis (as depicted in chapter 1, figure 4), being the phase *do* of greater influence thanks to the application of an analytical approach of this specific laser process.

Thus, a methodology for the description of a model begins from a theoretical analysis and ends in an experimental validation of analytical conclusions. Preceded by the methodology for the identification of influential parameters and its findings, the description of a model searches to form a reliable information basis for an efficient application of laser beam ablation with picosecond lasers in changing processing conditions and under continuous improvement. Figure 48 schematically illustrates a structural view of the analytical approach.

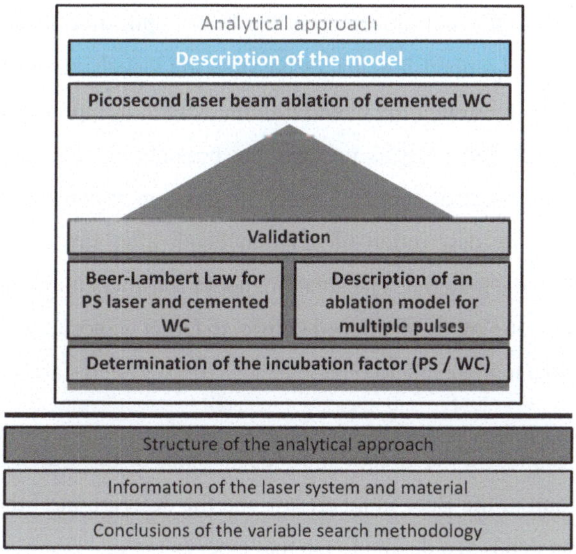

Figure 48: Structure of the analytical approaches for the description of the model

J. P. Calderón Urbina, *Efficient material laser beam ablation with a picosecond laser*, Light Engineering für die Praxis, https://doi.org/10.1007/978-3-662-61886-8_5

The core topics, in figure 48, are built with help of the findings of the variable search methodology and specific information of the properties of laser system and material, such as, thermal and optical properties of WC, and the model is achieved by the validation of the approaches. That is the reason why it is not only necessary to apply correctly the analytical approach but also to acquire and register system and material information to base the methodology on.

Specifically, the fifth chapter *Description of the model* pursues the understanding of picosecond laser beam ablation of cemented WC from an analytical point of view. First, the search of an incubation factor of cemented WC is analyzed from an experimental approach that quantifies the laser energy delivered on the material and the related posterior development of the material after different series of number of pulses. Posteriorly, a description of a model is targeted, and it is based on the knowledge of the behavior of electromagnetic waves through matter and adapted to the behavior of laser radiation and its specific ultra-short pulse laser beam characteristics in this work (e.g., picosecond pulse length, wavelength, absorptivity, etc.) through cemented tungsten carbide. This behavior is described by the Beer-Lambert law, which has been previously studied to some extent in pulse lengths, materials and processes in the laser application scope [99; 133]. The analysis with the Beer-Lambert law makes possible the visualization of the impact of a single picosecond pulse through the material in the length of approximately 10 ps; thus, an obtained two-dimension analysis presents the possible ablation depth and the heat regions based on the absorbed energy. Finally, an ablation model for multiple pulses is presented. It targets the description of ablation depth after the emission of multiple picosecond pulses on the material. The ablation model considers important characteristics of the laser and the material, as mentioned before, to target an accurate description of the behavior and to make it adaptable to other characteristics, systems and conditions.

The following lines present the already introduced topics of the search of the incubation factor, the adaptation of the Beer-Lambert law and the description of the ablation model in terms of ablation depth in a detailed systematic perspective.

5.1 Determination of the incubation factor

The effort to describe a model of a laser beam ablation process of a material must include as much information as possible to be closer to the real phenomenon; therefore, relevant particular material-related property values, such as, incubation and reflectivity, that are not available in the literature have to be determined. The search of these material properties in this study is done experimentally, and then deduced analytically. The determination of the incubation factor follows well known methods adapted to the interaction of cemented WC with a picosecond laser. Its importance impacts the understanding of multi-pulse laser beam ablation. Thus, the following lines describe a methodology for the determination of the incubation factor, its execution and the analysis of results.

5.1.1 Methodology

The determination of the incubation factor follows a hybrid experimental methodology based on known approaches, synthesized by the author and composed by two main parts; the first part, based on an approach for the measurement of Gaussian-beam spots diameters [134], helps to execute an accurate identification and determination of the threshold energy E_{th}, effective laser beam radius ω_0, and the respective threshold fluence F_{th} from the irradiation of a surface in a form of spots or tracks with a specific pulse energy E_0 (also as input energy or E_p) and number of pulses N. The second part, using the information of the first and applying the theory of equation 9 [103], focuses in the determination of the incubation factor for the studied material and laser. Figure 49 lists the steps of the methodology to determine the incubation factor.

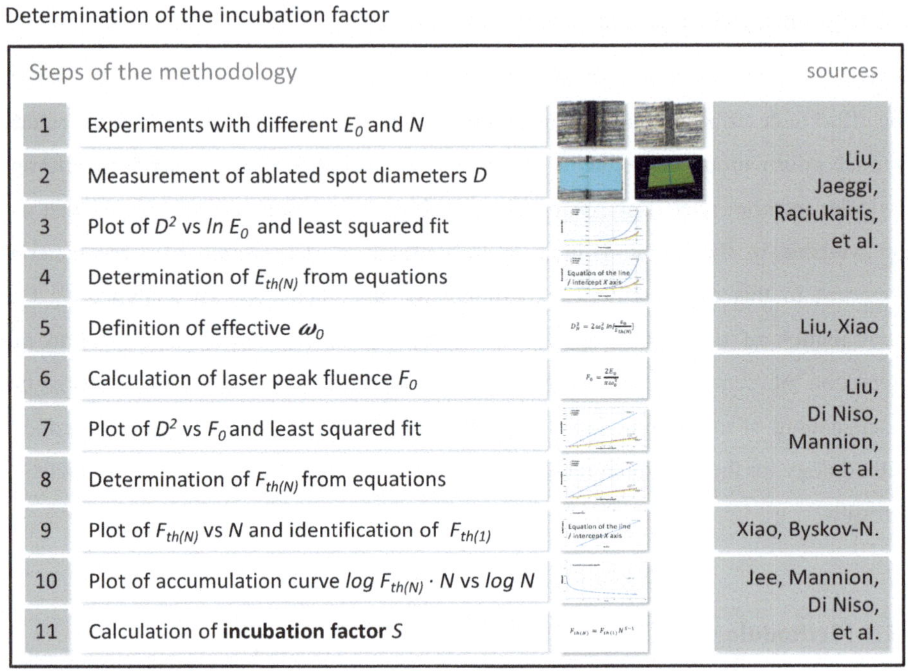

Figure 49: Steps of the methodology to determine the incubation factor.
Sources: [100; 101; 102; 103; 104; 106; 134; 135; 136; 137]

The ablation of the material starts when the input fluence of the source is put on the same level of the threshold fluence of the material. Thus, experiments with different input energies E_0 and number of pulses N have to be done to achieve craters or tracks of pulses (step 1) and gain contrast with the parameters and their levels. In the second step, a measurement of the diameter or width of the ablated spots or tracks D including heat affected zones (HAZ) is accomplished by the use of a confocal laser scanning microscope already described in subchapter 4.2.4 (Metrology) of this work with an average range of 200 measurement lines. These measurements help to analyze the relationship between input energy and material properties. The squared data of the measurement D^2, differentiated by number of pulses, is plotted in a graph in a ln function of E_0 in the third step. Moreover, a least-squares fit of the information is performed following the relation $D^2 \sim ln(E_0)$ to determine the energy threshold per number of pulses $E_{th(N)}$ from the equations of the lines and their respective ordinate-intercept to X axis (step 4). To calculate laser peak fluence, it is necessary to identify the effective beam waist radius ω_0; for this reason, the application of formula 21 [134; 104] facilitates this objective in step 5 by describing a relationship among the experiments of the first step and the interaction of the input and threshold energy; this relationship defines ω_0 by the slope ϕ^2 of the equation of the lines from the fourth step [134]. Following this method, a variable laser beam waist under certain limits is defined and compared when measurements are available. In the case of slight differences in the results after calculating the waists per number of pulses, an average can be obtained for comparison with the measurement and overall investigation of the incubation factor.

$$D_N^2 = 2\omega_0^2 \, ln\left[\frac{E_0}{E_{th(N)}}\right] \quad (21)$$

Once obtained the required information, in the sixth step laser peak fluence F_0 per number of pulses is calculated using equation 22 [101; 135; 136] and then plotted versus the squared diameters D^2 in the seventh step. The data is linear fitted and the respective equations of the resulting lines per number of pulses are obtained. Similar to step 4, the threshold fluence per number of pulses $F_{th(N)}$ is calculated from the equations of the lines and their ordinate-intercept (step 8). The resulting threshold fluences are plotted in function of the number of pulses in step 9 and from the logarithmic graph, the threshold fluence for one single pulse $F_{th(1)}$ is identified [104; 137].

$$F_0 = \frac{2E_0}{\pi\omega_0^2} \quad (22)$$

Finally, the required elements to determine the incubation factor are defined. Threshold fluences per number of pulses $F_{th(N)}$ are plotted as presented in figure 22 (B) (cp. Chapter 2) to obtain an accumulation curve by the logarithmic product of threshold fluences per number of pulses and number of pulses, $F_{th(N)} \cdot N$, in function of the logarithmic number of pulses N. The slope of this curve represents the variation of threshold fluence and hence S. The solution of equation 9 presents a numerical result of the incubation factor in the last step and it helps to have a point of reference of the information of the accumulation curve [103; 135].

5.1.2 Experimental determination of the incubation factor of cemented WC

The conditions of the experiments for the investigation of the incubation factor were based on the conditions of the available laser system described in chapter 4 (under 4.2.1 System equipment). The system determines the restrictions of pulse repetition rate and scanner speed, which influences the isolation of pulses and defines a specific number of them. The number of pulses were defined following equation 15 of pulse overlap (chapter 4). The laser beam power P_L was varied from 10% to 100% in parts of 10% to target a variation of input energy E_0. The general conditions of the experiment can be seen on table 14. In the first step of the methodology, ablation experiments of different input energies and three specific number of pulses (N = 3, 28 and 224 pulses) were set and irradiated in form of tracks (laser ablated lines). These levels of N were selected according to the available system capabilities and limits (pulse repetition rate and scanner speed) and targeting different orders of magnitude.

Table 14: Parameters and levels

Parameter	Remarks	Levels	Units
Laser beam power P_L	P_L = 50 W at 100%	10, 20, 30, 40, 50, 60, 70, 80, 90, 100	[%]
Number of pulses N	At f = 400 kHz,	3, 28, 228	[-]
Pulse overlap PO	v_s = 3730, 400, 50 mm/s (eq. 12)	67, 96, 100	[%]
Focus position z_f	Position: focus vs material	0	[mm]
Incidence angle γ	Laser beam to workpiece	90	[°]
Material	A plate of cemented WC (as described in subchapter 4.2.2)		

Figures 50, 51 and 52 present the sets of experiments (from 1 to 10 tracks) by number of pulses N = 3, N = 28 and N = 224, respectively. In the figures, it can be observed the development of the raise of the laser beam power P_L, i.e., input energy E_0 at f = 400 kHz, by an increment of 10% of the power by each subsequent experiment and until P_L = 100% is reached. The influence of the increase of laser power can be observed in every posterior track by the width of the track and the corresponding HAZ. The samples of cemented WC were only cleaned and not chemically or mechanically prepared or polished; it was intended to maintain possible industrial processing conditions of the samples.

Once the tracks were ablated under the selected conditions, measurements of the width of the tracks were done and defined as D. The results of the measurement of the three sets of experiments and the calculation of the corresponding pulse energy for the conditions of the experimentation are also presented in figures 50, 51 and 52.

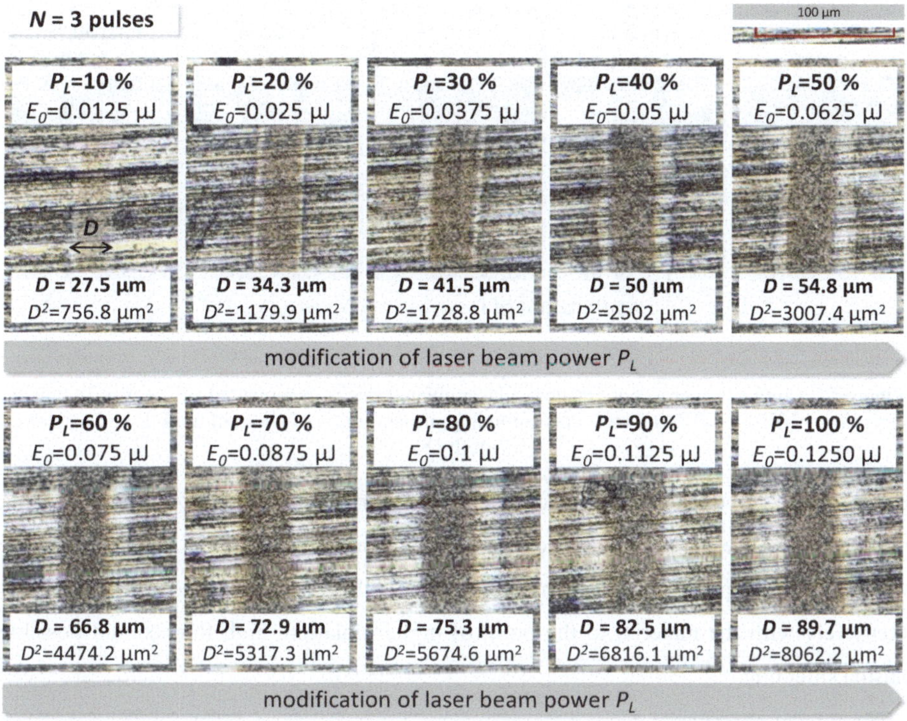

Figure 50: Confocal microscope images of laser ablated tracks in cemented WC at $N = 3$ pulses from $P_L = 10\%$ to 100% and their corresponding measurement of width of track D and squared width D^2

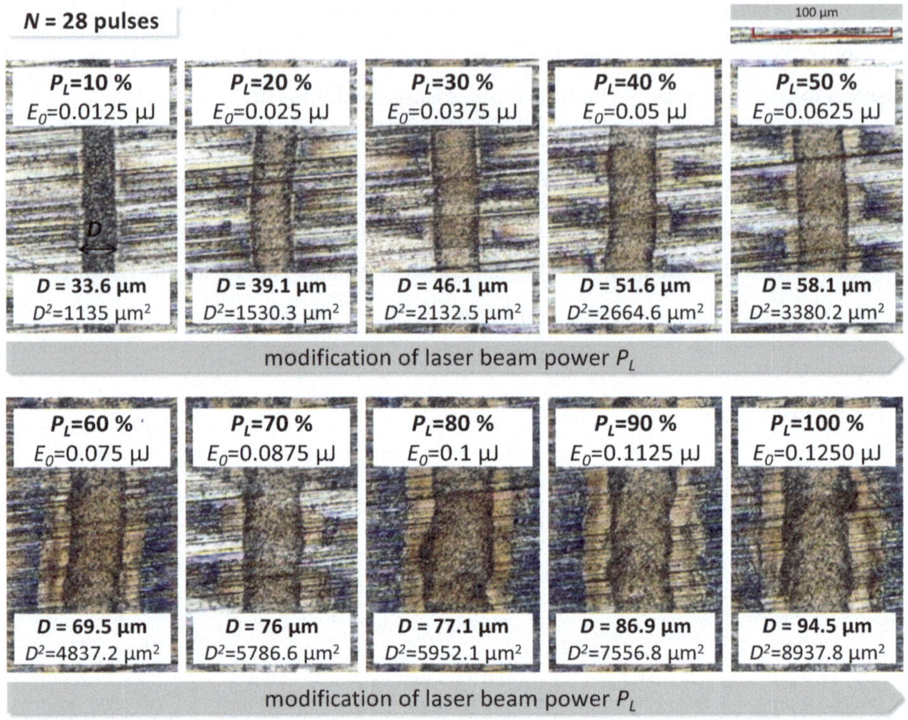

Figure 51: Confocal microscope images of laser ablated tracks in cemented WC at $N = 28$ pulses from $P_L = 10\%$ to 100% and their corresponding measurement of width of track D and squared width D^2.

The influence of laser beam power can be easily appreciated in $N = 224$, where a higher ablation can be measured in widths of the tracks and a darker color can be observed. The high number of pulses, when compared to the other sets of experiments, interacts intensively with the material to the point of higher ablation, high fluence and possible plasma shielding.

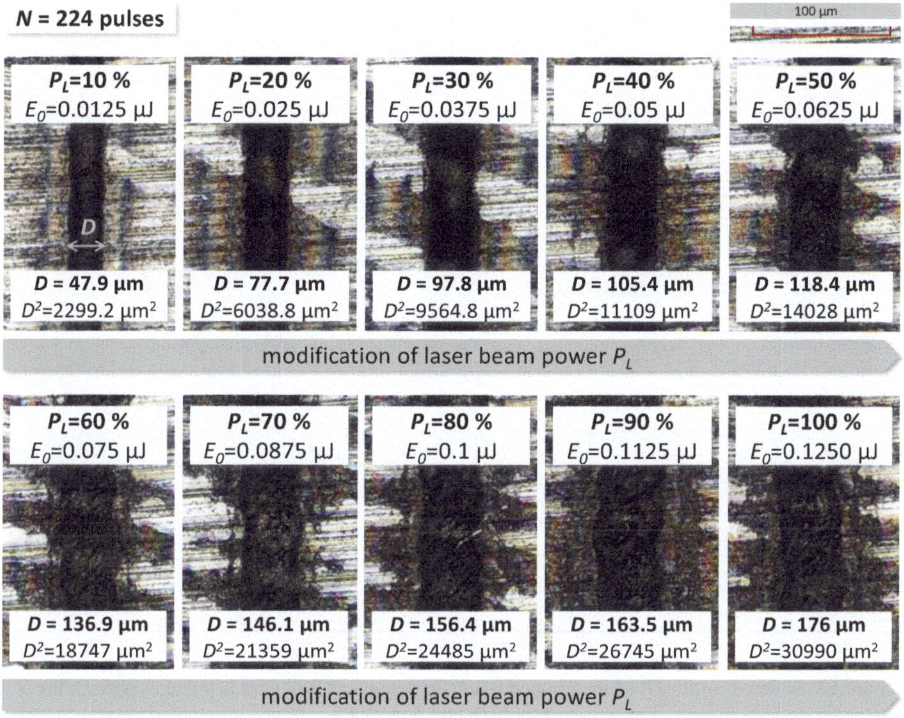

Figure 52: Confocal microscope images of laser ablated tracks in cemented WC at $N = 224$ pulses from $P_L = 10\%$ to 100% and their corresponding measurement of width of track D and squared width D^2.

The squared width of the tracks D^2, also presented in figures 50, 51 and 52, represents an important piece of information for the semi log plot of D^2 versus E_0. Based on the plot, the data is least-squares fitted accordingly to $D^2 \sim ln(E_0)$ and the results are shown in figure 53. The three resulting curves corresponding to the E_0 of each set of N's are prolonged to their intercept to X axis, and thus being able to determine their threshold energy $E_{th(N)}$.

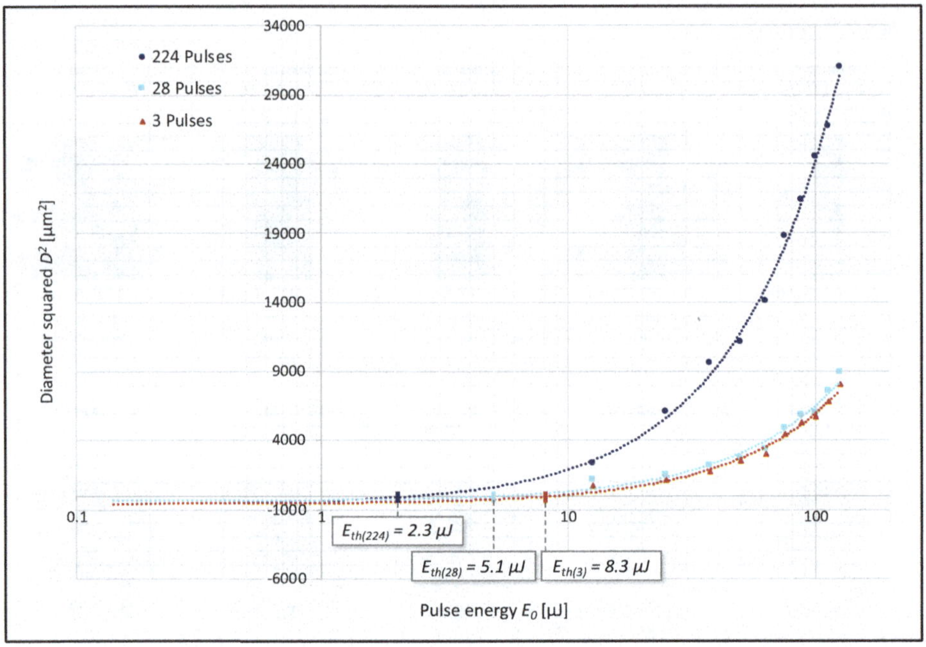

Figure 53: Squared width D^2 of the ablated tracks in cemented WC versus pulse energy E_0 for N = 3, N = 28 and N = 224

The resulting energy thresholds for $N = 3$, $N = 28$ and $N = 224$ are $E_{th(3)} = 8.3$ µJ, $E_{th(28)} = 5.1$ µJ and $E_{th(224)} = 2.3$ µJ, respectively. These results, in combination with beam waist radius ω_0 and according to equation 22, contributes to the determination of the laser peak fluence F_0 and consequently of the threshold fluence per number of pulses $F_{th(N)}$. Following the methodology, the slope of the curves hints a beam waist radius range of ω_0 = 8 to 15 µm that confirms the characterization of the picosecond laser beam at $f = 400$ kHz in chapter 4 of ω_0 = 14 µm. This measurement by optical tools is defined as the effective beam waist radius ω_0 for the calculation of the laser peak fluence. The application of equation 22 concludes in the determination of laser peak fluence F_0 per number of pulses and results in three series of data illustrated in figure 54.

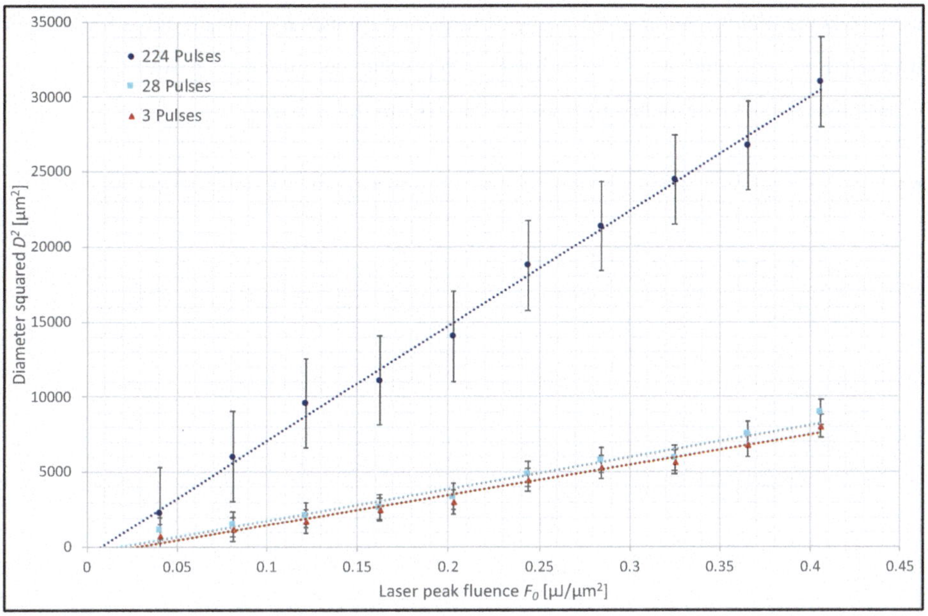

Figure 54: Squared width D^2 the ablated tracks in cemented WC versus laser peak fluence F_0 for $N = 3$, $N = 28$ and $N = 224$

The data is least-squares fitted and three lines corresponding to N are obtained. From these lines, it is possible to identify their threshold fluence $F_{th(N)}$, the minimum energy per area to start ablating the material, by intercepting them to the X axis. The results are presented in the following table (table 15).

Table 15: Resulting threshold fluence $F_{th(N)}$ for cemented WC

Number of pulses N	Threshold fluence $F_{th(N)}$	
[-]	[µJ/µm²]	[J/cm²]]
3	0.027	2.69
28	0.016	1.66
224	0.007	0.74

The plot of threshold fluence per number of pulses $F_{th(N)}$ versus number of pulses N in figure 55 illustrates the development of threshold fluence by increasing the number of pulses. Moreover, a least-squares fit of the data and its logarithmic projection to $x = 0$

simplifies the identification of the value of threshold fluence $F_{th(1)}$ when only one pulse is irradiated on cemented WC. This value $F_{th(1)}$ represents an important part of the solution of equation 9 and corresponds to $F_{th(1)} = 0.032\ \mu J/\mu m^2$ ($3.2\ J/cm^2$).

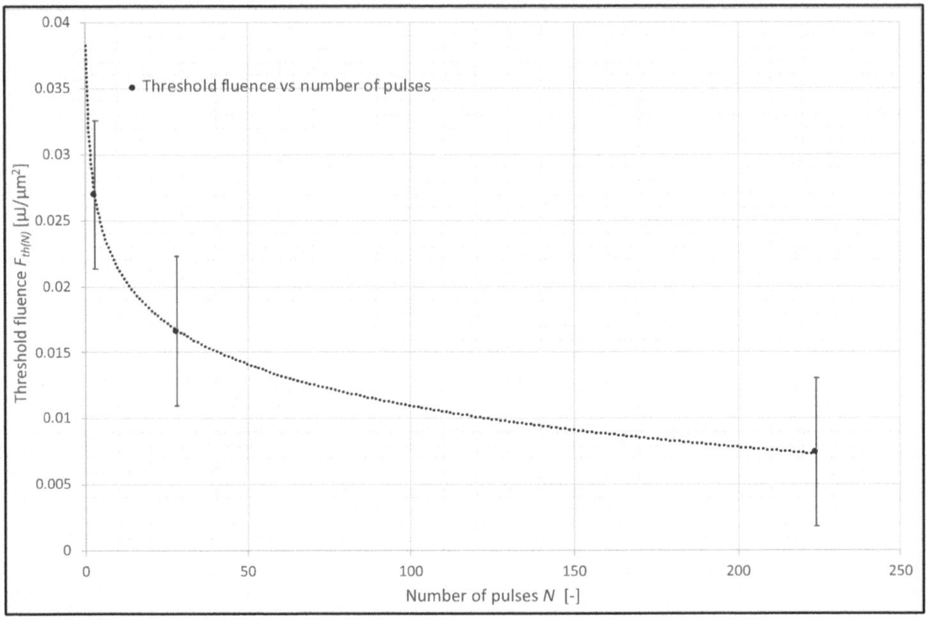

Figure 55: Development of threshold fluence $F_{th(N)}$ as a function of the number of applied pulses ($N = 3$, $N = 28$ and $N = 224$ pulses) for cemented WC

The reduction of threshold fluence, observed in figure 55, describes the behavior of the material properties when the number of applied pulses is modified. Thus, the adaptation of this behavior as an accumulation curve defines the factor of accumulation, i.e., incubation factor. Figure 56 illustrates the accumulation curve for cemented WC by the logarithmic product of threshold fluence per number of pulses and number of pulses, $F_{th(N)} \cdot N$, versus the logarithmic number of pulses, N.

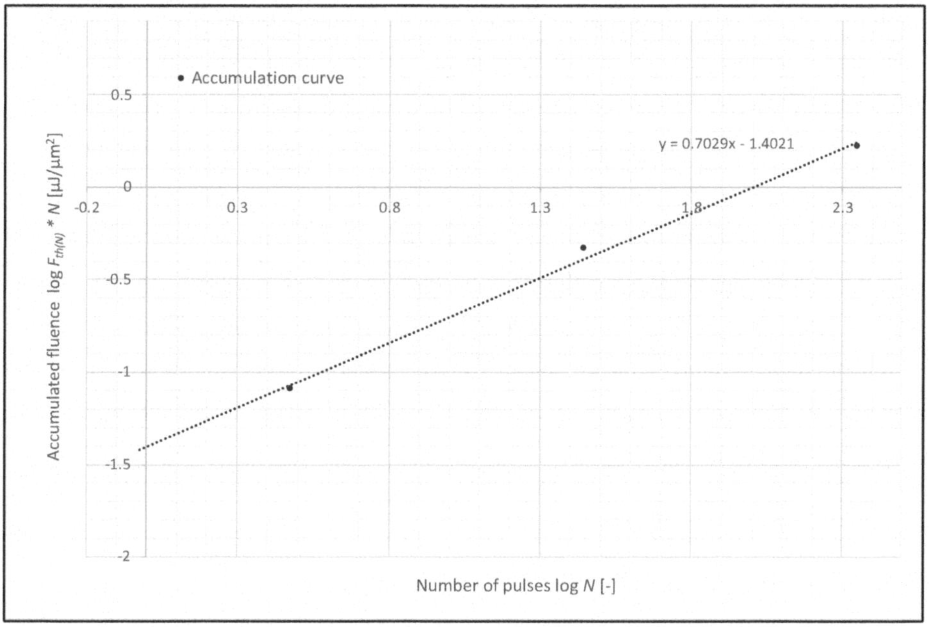

Figure 56: Accumulation curve for cemented WC sample for $\tau_H = 10$ ps and at $f = 400$ kHz. The data is least-squares fitted according to equation 9 with a $R^2 = 0.99$

The data is least-squares fitted and the incubation factor is defined by the slope of the resulting line. Therefore, the incubation factor for cemented WC irradiated with a picosecond laser $\tau_H = 10$ ps, $\lambda = 1064$ nm and at $f = 400$ kHz resulted in this methodology of $S = 0.70$. The solution of equation 9 for each number of pulses N gives for $N = 3$, a $S = 0.84$, for $N = 28$, a $S = 0.80$, and for $N = 224$, a $S = 0.73$, respectively.

5.1.3 Conclusion

The determination of the incubation factor is a process of careful control and analysis of information and experimentation. In industrial environments, the adaptation of theory to industrial set-ups represents a challenge for the accurate application of methods. Specifically, the separation and isolation of certain number of pulses might require the adequate configuration of all system elements. The determination of the incubation factor for this work guided the investigation from the design of the experimentation, through the analysis of the output data. Previous experiences in the matter supported the adaptation to the particular system conditions of this work, e.g., picosecond laser, cemented WC, number of pulses, etc. Finally, an incubation factor for cemented WC was found from the linear representation of the accumulation curve and also from the solution of the equation proposed by Jee; both results specify an incubation factor lying in the range of $\boldsymbol{S} = \mathbf{0.7\text{-}0.8}$ ($S = 0.75$ for precise use). This information is used in the following parts of this work to analyze laser beam ablation of cemented WC with a picosecond laser. The investigation of a *variable incubation factor* for a same material under different conditions of laser and system, e.g., pulse length, fluence, wavelength, number of pulses, among others, has to be considered for further studies.

5.2 Determination of the absorptivity factor

The investigation of the absorptivity factor in this work is as import as the determination of the incubation factor for the understanding of the laser beam ablation process on cemented WC. The analysis requires as much information of the material and system as possible to describe an accurate picosecond laser beam ablation model. The lack of information in the literature of compound and novel materials and their characteristics, e.g., optical properties, is constant. For this reason, an experimental approach to identify the absorptivity factor at infrared light of a blank of cemented WC is conducted.

An experimental set-up was configured at the *Deutsches Elektronen-Synchrotron*, DESY Hamburg, with the basic idea of irradiating a beam of infrared light with an average beam power P_L to a polished sample of cemented WC and capturing its reflection by means of a power measurement device at different emission angles. A polished laser additive manufactured tool steel sample was also irradiated to have a reference measurement. First, the samples were prepared on a standard mount and embedded in a polymer matrix for grinding and polishing. The polishing process was done in a Streuers Tegramin-30 grinding and polishing machine to obtain a mirror-like aspect.

Second, the experimental set-up was prepared with three elements. First, an infrared laser beam source, Crystal Laser CL-200 diode pumped, at λ = 1064 nm and emitting a measurement power of P_L = 225 mW; second, a power measurement device, Coherent LabMax-TOP, and third, the measurements samples; all components set on a standard precision laboratory table. The elements of the configuration are presented in figure 57 before they are placed in its final measurement position.

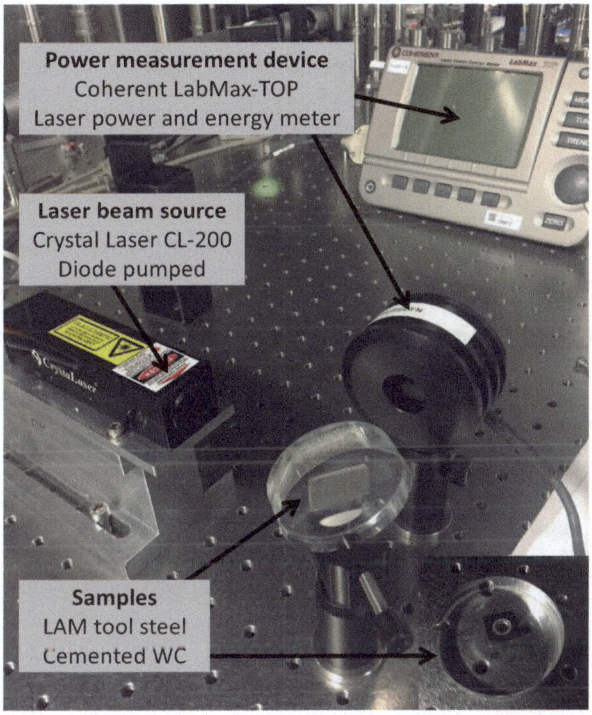

Figure 57: Elements of the configuration for the measurement

The set-up starts with the emission of the beam from the source to the sample, which is positioned in an angle θ that redirects the beam to the measurement device to capture it after its reflection on the sample; thus, the elements of the configuration form a triangle between them. Moreover, three acute angles were defined to analyze the reflection. The first angle is intended to be the closest to zero degrees and due to the size of the components, the restrictions of the configuration and the possibility of an effective measurement of the beam, it was set to 1.55°; the second angle was set at 7°, and the third was set at 45°. The determination of the angles was executed by setting the distances between the elements and defining their trigonometric relationships; the identification and calibration of the position of the beam was done by the use of an infrared beam detector and viewer.

The objective of the measurement is to obtain a reflectivity factor that corresponds to the angle 0° or that is the closest to this perpendicular position of the laser beam to the irradiated surface as in the case of laser material processing (closest angle to 0° achieved in this work: 1.55°).

A schematic representation of the final configuration of the experimental set-up for the measurement of reflectivity (reflectance) can be observed in figure 58. It can be seen that the distance between the laser beam source and the samples was kept constant to 120 cm and from this, it was set the distance between the power measurement device and the source to get the selected reflection angle θ as a tangent function.

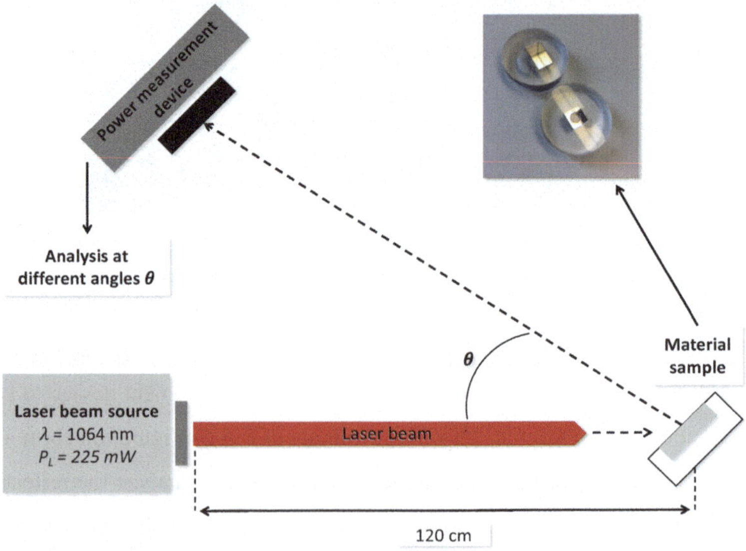

Figure 58: Schematic representation of the experimental set-up for the measurement of reflectivity

Once the set-up is ready, a preliminary measurement of the reflectivity is done at the same angle but at different positions on the sample to discard irregularities of the measurement, homogeneity or surface of the piece (for this study four positions and the measurements gave no irregularities or relevant discrepancies for both samples). After this, the measurement is executed by one sample at each angle, and then, the other sample at each angle, in a similar manner. The measured laser beam power for all samples and angles was registered and then contrasted to the emitted laser beam power to determine the reflectivity factor *R* (equation 23) that is needed for the calculation of the absorptivity factor, based on equation 24.

$$R = \frac{P_{L,R}}{P_{L,0}} \tag{23}$$

Where, $P_{L,0}$ is the initial laser beam power emitted from the source; and, $P_{L,R}$ corresponds to the laser beam power reflected from the sample and captured by the measurement device.

Equation 24 describes the complementary relationship of reflectivity and absorptivity when an optical radiation is emitted on a material [138],

$$R + A = 1 \tag{24}$$

Where, *A* is the absorptivity factor, and it is complemented by *R* to equal 1. Transmittance *T* is included in this expression when the irradiation is done on transparent materials, which is not the case of cemented WC and tool steel, materials that are considered opaque.

The results of the measurements and the calculation of reflectivity and absorptivity factors are presented in table 16.

Table 16: Results of the measurement and conclusions

Angle of reflection	Laser beam power	Cemented WC factors		Laser beam power	LAM tool steel factors	
θ	$P_{L,R}$	Reflectivity R	Absorptivity A	$P_{L,R}$	Reflectivity R	Absorptivity A
[°]	[mW]	[-]	[-]	[mW]	[-]	[-]
1.55	76	0.33	0.67	150	0.66	0.34
7	76	0.33	0.67	149	0.66	0.34
45	55	0.24	0.76	136	0.60	0.40
$P_{L,0}$ = 225 mW						

As observed in table 16, the determination of the absorptivity factor for LAM tool steel in this study corresponds to the factor found in the literature for steel or stainless steel; this confirms the validity of the method [139]. Therefore, the given results for cemented WC are taken as representative. In conclusion, an experimental method with a reference material was done to obtain the reflectivity factor of cemented WC when irradiated with an infrared laser beam. The investigation gave a reflectivity factor value of $\boldsymbol{R}$ **= 0.33**, i.e., an absorptivity factor value of 0.67, which are accepted and included in the analysis of the model in the following part of this work. Similarly, a relevant discussion of the influence of reflectivity in the process of ultra-short pulse laser ablation by some authors is discussed in the next part of this thesis.

5.3 Adaptation of the Beer-Lambert law to picosecond pulse laser ablation

The fundamental idea of laser beam ablation is based on the fact that sufficient energy has to be transferred to the material to be able to evaporate it or sublimate it. In the case of picosecond pulses, this transfer of energy is dependent of their pulse length that interacts in a short period of time with the material following the characteristics of the ultra-short pulse regime. Therefore, in order to gain understanding of this phenomenon, the adaptation of the Beer-Lambert law to picosecond laser ablation of a single pulse on cemented WC is applied in the following part of the study.

5.3.1 Theoretical background

The description of the Beer-Lambert law of linear absorption of an incident electromagnetic radiation (e.g., a picosecond laser pulse) moving through a material (e.g., cemented WC) is reduced exponentially with the thickness of the material for a given wavelength [24; 99]. Therefore, the distribution of the pulse energy through the material facilitates the understanding of two relevant aspects; first, the identification of the energy limits for ablation, i.e., the delimitation of threshold fluence F_{th}; and second, the behavior of an irradiated laser pulse as the development of average input laser fluence F through time, in terms of intensity. Both aspects together define the depth of the ablation in the material and the delimitation of other thermal zones.

A fundamental approach of the Beer-Lambert law, as presented in equation 4 in chapter 2, includes the intensity of the radiation and the magnitude of length when travelling through the material. The adaptation proposed in this chapter incorporates the concepts of heat and time of the irradiation [99]. Thus, equation 25 expresses the concept of the Beer-Lambert law for laser processing.

$$Gt(y,t) = I(t) \cdot (1 - R(T_r)) \frac{e^{\frac{-y}{\delta}}}{\delta} \tag{25}$$

Where, $Gt(y,t)$ corresponds to the heat source in depth y and time t, $I(t)$ is the laser pulse intensity through time, R is the reflectivity factor at a temperature T_r, and δ is the optical penetration depth.

The concept of optical penetration depth δ is a fundamental aspect in the transmission of heat as shown in equation 25 and describes how profound light or any electromagnetic emission can penetrate into the material. The beaming of an incident electromagnetic radiation on a surface results in the partial reflection of the beam, based on its reflectivity factor, and the partial absorption of the beam energy into the material. According to the material and its properties, the interaction of the electromagnetic radiation with electrons and lattice reaches a specific depth defined by δ and related to the absorption coefficient α. Equation 26 defines α as a function of the wavelength of the beam:

$$\alpha = \frac{4 \cdot \pi \cdot \kappa}{\lambda} \quad (26)$$

with λ representing the wavelength of the electromagnetic radiation or incident laser beam and κ the absorption index of the material.

Moreover, penetration depth δ and absorption coefficient α of a material maintain an inversely proportional relationship as stated in equation 27 and from which δ can be expressed as a function of the wavelength of the beam λ from equation 26 and 27 as stated in equation 28.

$$\delta = \frac{1}{\alpha} \quad (27)$$

$$\delta = \frac{\lambda}{4 \cdot \pi \cdot \kappa} \quad (28)$$

An important topic for the understanding of the interaction of the laser beam with the material and its thermal effects in the ablation are the basic heat transfer equations [45; 99; 140]. These concepts describe two ideas; first, the required power to rise the temperature of a volume of the target material, and second, the required power to change the physical phase of that volume of target material. From this understanding, it is possible to identify the required power per cubic meter to evaporate a solid material, and consequently, define theoretically its threshold fluence from the information of the material properties and the laser beam specifications. Equations 29 and 30 introduces the basic heat transfer expressions to this work:

$$\varepsilon_v = c_p \cdot \rho \cdot \frac{dT}{dt} \quad (29)$$

In (29), ε_v represents the energy (also as power density for this work), required to increase the temperature of the material, c_p is the specific heat capacity of the material, ρ

corresponds to the density of the material, dT is the change in the temperature and dt is the time period for the change in the temperature to happen.

Equation 30 describes the energy ε_{υ} needed to change the phase of the material from whether solid to liquid or liquid to vapor:

$$\varepsilon_{\upsilon} = \frac{LH \cdot \rho}{t} \tag{30}$$

with LH representing the latent heat related to the change of phase of the target material.

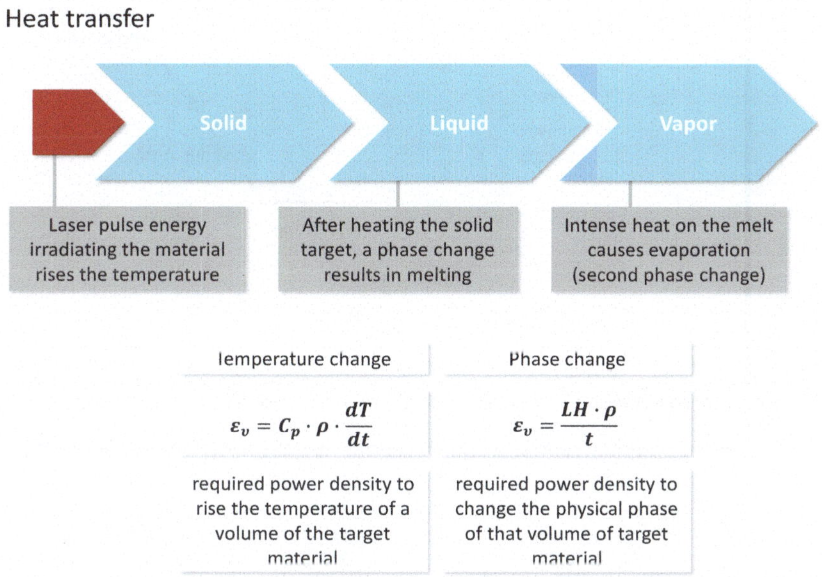

Figure 59: Heat transfer fundaments

Figure 59 schematically explains the basic idea of heat transfer into the material. A laser beam transmits its energy into the surface of the target material and heats it. Then, a rise on the temperature causes a phase change that melts the solid. The continuous heat on the melted zone increases the temperature until the liquid changes into vapor by a second phase change. Thus, an ablation of the solid is done. In the case of ultra-short pulse ablation for this subchapter 5.3, this heat transfer process and phase changes happen in the range of the τ_H of ps pulses.

Temperature increase and phase change can be observed in figure 60. It illustrates the path of a heated material as a function of energy. The rise of sufficient temperature as a consequence of energy irradiation causes heating and phase change that results in the ablation of the target material.

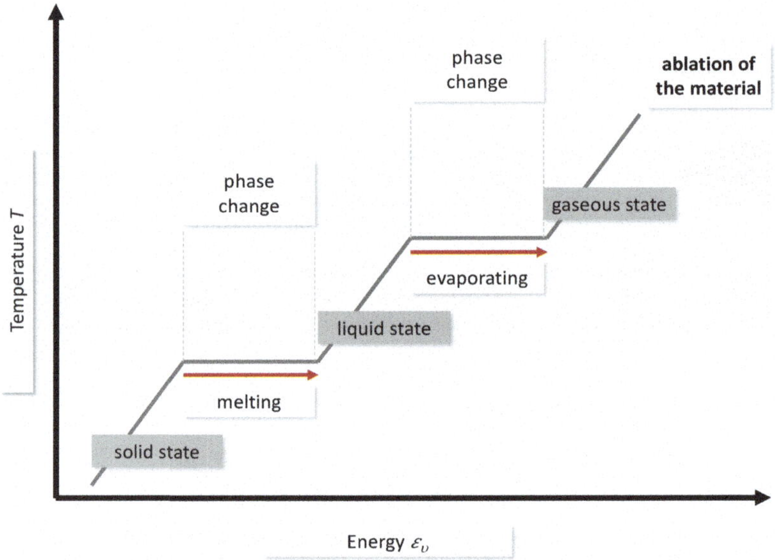

Figure 60: Phase change diagram [141]

The calculation of the required total power density $\varepsilon_{\upsilon T}$ to ablate, i.e., evaporate, a volume using equations 29 and 30 leads to the determination of the threshold fluence of the material in function of pulse length and penetration depth, as stated in equation 31.

$$F_{th} = \varepsilon_{\upsilon T} \cdot \tau_H \cdot \delta \tag{31}$$

The threshold fluence sets the borders of ablation and defines the different affectation zones of the input energy from the laser source into the material. Based on this knowledge, it is important to define the input laser energy to identify the penetration of the input power. Thus, the energy distribution of the pulse follows, according to its characterization, a gaussian behavior through time and it is expressed in equation 32 as Gaussian input power intensity P_I after [99]. For the application of the Beer-Lambert law, P_I corresponds to laser pulse intensity through time $I(t)$.

$$P_I = \frac{E_0}{\pi \cdot \omega_0^2 \cdot \tau_H} e^{\frac{4(\Delta t - \tau_{Hm})^2}{\tau_{Hm}{}^2}} \tag{32}$$

where E_0 is the pulse energy and τ_H is the pulse length as mentioned before in this work; $\pi \cdot \omega_0^2$ represents the area of the laser spot, τ_{Hm} corresponds to the half of the pulse length or $\tau_H/2$, and Δt represents a period of time of τ_H.

The control of the Gaussian input power intensity P_I facilitates the irradiation of a targeted amount of heat from the source in terms of power density on a volume Gt. The identification of the threshold power density of the material $\varepsilon_{\upsilon T}$, i.e., threshold fluence F_{th}, thanks to the heat transfer equation 29 and 30, guides the application of both mentioned concepts to ablate the material. The definition of average input laser fluence of the system F, described in equation 33, is based on equation 32 and includes the characteristics and the concept of a gaussian laser pulse profile and the Beer-Lambert law [99], respectively.

$$F = \tau_H \cdot \frac{\sum_{i=1}^{m} \frac{E_0}{\pi \cdot \omega_0^2 \cdot \tau_H} e^{\frac{4(\Delta t_i - \tau_{Hm})^2}{\tau_{Hm}{}^2}}}{1 + 2 + \cdots + m} \tag{33}$$

Finally, all the required elements to describe the behavior of a Gaussian laser beam pulse into the material are defined. Specifically, the threshold fluence deduced from the material properties is understood and the power intensity to ablate the material is defined. Together both concepts enable the visualization of the pulse distribution through the material per time, volume and energy (power density), and define the zone of the ablated material by fulfilling the elements of the Beer-Lambert law in equation 25.

5.3.2 Adaptation to picosecond pulse laser beam ablation of cemented WC

Based on the theoretical background, a procedure to execute an adaptation of the Beer-Lambert law to picosecond pulse laser beam ablation of cemented WC is conceived. Figure 61 describes the procedure, which starts with the collection and identification of relevant information and specifications from material and laser system for the analysis.

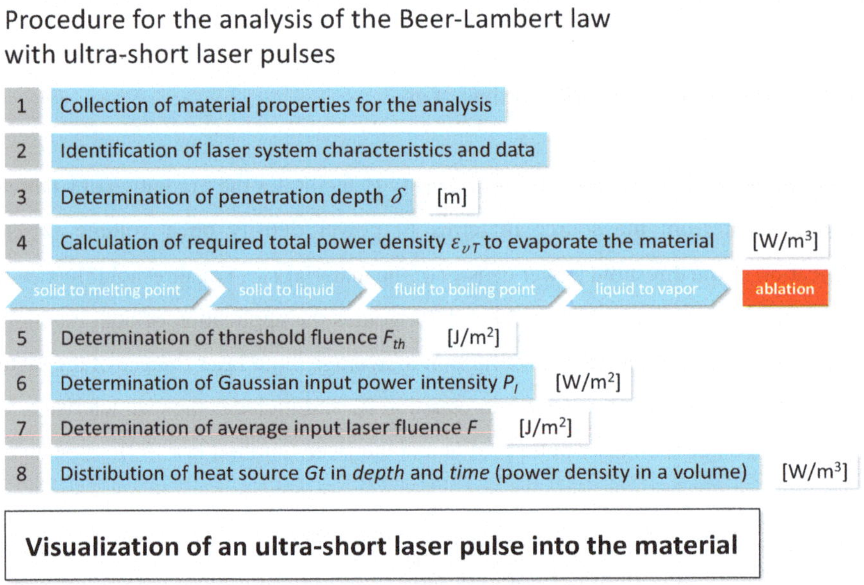

Figure 61: Schema for the adaptation of the Beer-Lambert law to picosecond pulse laser beam ablation of cemented WC

Then, the penetration depth for the laser into the material is determined. Based on the heat transfer theory, a calculation for the identification of the total power density to ablate the material is executed; from this calculation, the threshold fluence for the material is identified. The laser information contributes with the determination of the Gaussian input power intensity and its related average input laser fluence. Finally, the procedure concludes with the presentation of the distribution of the heat source in depth and time in the material and facilitates the understanding and visualization of the laser pulse through the material.

The summary of relevant characteristics of cemented WC is listed in table 17 and contains important optical and thermal properties for the for the analysis. Characteristics of the laser system, necessary for the breakdown of the method, are already mentioned in subchapter 4.2.1.

Table 17: Material properties and laser system characteristics for the analysis

Properties / parameters	Element	Value	Unit
Thermal conductivity k	WC	164	$W\ m^{-1}\ K^{-1}$
Specific heat (solid-state) $C_{p,s}$	WC	210	$J\ kg^{-1}\ K^{-1}$
Specific heat (liquid-state) $C_{p,l}$	WC	292 ♣	$J\ kg^{-1}\ K^{-1}$
Mass density (solid-state) ρ_s	WC	15930	$kg\ m^{-3}$
Mass density (liquid-state) ρ_l	WC	14500 ♣	$kg\ m^{-3}$
Melting point T_m	WC	3143	K
Boiling point T_b	WC	6273	K
Latent heat of fusion LH_f	WC	400 000	$J\ kg^{-1}$
Latent heat of evaporation LH_e	WC	4 020 000	$J\ kg^{-1}$
Absorption index κ	WC / W	3.77 (value for W)	-
Absorption coefficient α	WC	4.45×10^{7} ♥	m^{-1}
Incubation factor S	WC	0.75 ♠	-
Reflectivity R	WC	0.33 ♦	-

♣ No available, based on solid-state value
♥ From equation 26
♠ Determined in subchapter 5.1
♦ Determined in subchapter 5.2

Sources: [116; 117; 110; 118; 119; 120; 121]

Once presented the material and laser system information, it is necessary to determine the laser ablation borders of the material to understand and control how much energy has to be systematically irradiated to ablate the material. For this, it is necessary to identify the ablation threshold of cemented tungsten carbide based on two important aspects: optical penetration depth and the required total power density to evaporate the material.

Therefore, the optical penetration depth of the laser into the material has to be determined, using equation 26 by targeting the absorption coefficient of the material,

$$\alpha = \frac{4 \cdot \pi \cdot 3.77}{1.064 \, x \, 10^{-6} \text{ m}} \quad (34)$$

The resulting absorption coefficient $\alpha = 4.45 \text{ x } 10^{7} \text{ m}^{-1}$ is used in equation 27 to find its corresponding inversely proportional factor: the penetration depth δ. Thus, it is calculated for cemented WC under infrared irradiation of $\lambda = 1064$ nm an optical penetration depth of $\delta = 2.25 \text{ x } 10^{-8}$ m.

Knowing the value of δ, the procedure continues with the calculation of the required total power density to evaporate cemented WC. Heat transfer basic equations 29 and 30 are required to solve the process from the solid sample to the evaporation. The basic idea is how much energy is required to evaporate a volume of material with a 10 ps pulse and accomplishing the temperature and phase changes accordingly. Therefore, the analysis includes the calculation of the required energy to rise the temperature to the liquidation boundary and the energy to change the phase of WC from solid to liquid; similarly, it is needed to calculate the required energy to increase the temperature from liquid to the boiling point and the energy to change the phase from fluid WC to vapor [8]. Figure 62 illustrates this idea. The total power density $\varepsilon_{\upsilon T}$ to evaporate solid cemented WC is the result of the addition of each required value of power density ε_{υ} to both change the temperature and the phase of the material from solid-state to vapor-state, passing through liquid-state. The energy transfer occurs in picosecond time-range; hence, heat losses caused by conduction or any other conditions are neglected.

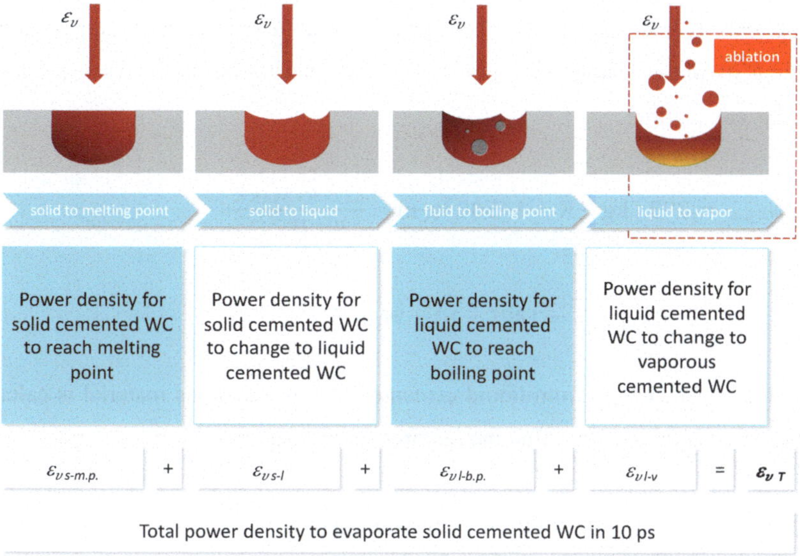

Figure 62: Temperature and phases change for cemented WC

The required power density to change the temperature of solid cemented WC at a room temperature of 293 K and reach its melting point is described by equation 29:

$$\varepsilon_{v\ s-m.p.} = 210\ \mathrm{J\ kg^{-1}\ K^{-1}} \cdot 15930\ \mathrm{kg\ m^{-3}} \cdot \frac{3143\ \mathrm{K} - 293\ \mathrm{K}}{10\ \mathrm{x}\ 10^{-12}\ \mathrm{s}} \tag{35}$$

$$\varepsilon_{v\ s-m.p.} = 9534\ \mathrm{x}\ 10^{20}\ \mathrm{W\ m^{-3}} \tag{36}$$

The necessary power density to reach a phase change from solid to liquid cemented WC is calculated with equation 30 and presented below:

$$\varepsilon_{v\ s-l} = \frac{400000\ \mathrm{J\ K^{-1}} \cdot 15930\ \mathrm{kg\ m^{-3}} \cdot}{10\ \mathrm{x}\ 10^{-12}\ \mathrm{s}} \tag{37}$$

$$\varepsilon_{v\ s-l} = 6372\ \mathrm{x}\ 10^{20}\ \mathrm{W\ m^{-3}} \tag{38}$$

Once the cemented WC has been calculated to have a liquid-state, it is necessary to obtain the power density that is required to change its temperature to reach its boiling point before calculating its phase change to vapor-state. Following the same analysis, equation 29 guides this calculation:

$$\varepsilon_{v\,l-b.p.} = 292\ \mathrm{J\,kg^{-1}\,K^{-1}} \cdot 14500\ \mathrm{kg\,m^{-3}} \cdot \frac{6273\ \mathrm{K} - 3143\ \mathrm{K}}{10 \times 10^{-12}\ \mathrm{s}} \tag{39}$$

$$\varepsilon_{v\,l-b.p.} = 1325 \times 10^{21}\ \mathrm{W\,m^{-3}} \tag{40}$$

Finally, the phase change from liquid cemented WC to vaporous material is calculated with the use of equation 30 to find the required power density:

$$\varepsilon_{v\,l-v} = \frac{4020000\ \mathrm{J\,K^{-1}} \cdot 14500\ \mathrm{kg\,m^{-3}} \cdot}{10 \times 10^{-12}\ \mathrm{s}} \tag{41}$$

$$\varepsilon_{v\,l-v} = 5829 \times 10^{21}\ \mathrm{W\,m^{-3}} \tag{42}$$

Having obtained all partial power densities to transform solid cemented WC to vapor, it is possible to determine the total volume power density $\varepsilon_{v\,T}$ by adding the results of expressions 36, 38, 40 and 42:

$$\varepsilon_{v\,T} = 874485 \times 10^{21}\ \mathrm{W\,m^{-3}} \tag{43}$$

The identification of the total volume power density to ablate the material enables the identification of its threshold fluence. This information makes possible to target a defined specific range of energy to ablate the material. Equation 31 defines the threshold fluence for cemented WC in this study with the use of penetration depth and pulse length:

$$F_{th} = 874485 \times 10^{21}\ \mathrm{W\,m^{-3}} \cdot 10 \times 10^{-12}\ \mathrm{s} \cdot 2.25 \times 10^{-8}\ \mathrm{m} \tag{44}$$

$$F_{th} = 1960\ \mathrm{J\,m^{-2}} \tag{45}$$

The value of $F_{th} = 1960$ J m^{-2} ($F_{th} = 0.1964$ J cm^{-2}) represents the total energy per area for a 10 ps pulse to evaporate solid cemented WC from the perspective of the thermal and optical properties of the material.

Once defined the ablation threshold, the following step in the procedure is to determine the input power intensity of the system adapted to a Gaussian beam profile. Both concepts ablation threshold and input power intensity enable the understanding and visualization of the picosecond pulse through the material. The calculation of the Gaussian input power intensity P_I, described by equation 32, represents $I(t)$ in equation 25 and is determined by periods of time Δt, which corresponds to segments of the pulse length. This partition allows the understanding of the Gaussian path of the pulse through the material in terms of power density, depth and time, as it is posteriorly presented in this work. The periods Δt for this analysis are set to $\Delta t = 0.5$ ps. Figure 63 illustrates the time distribution of a picosecond pulse in terms of power intensity through time and confirms the Gaussian behavior of the intensity.

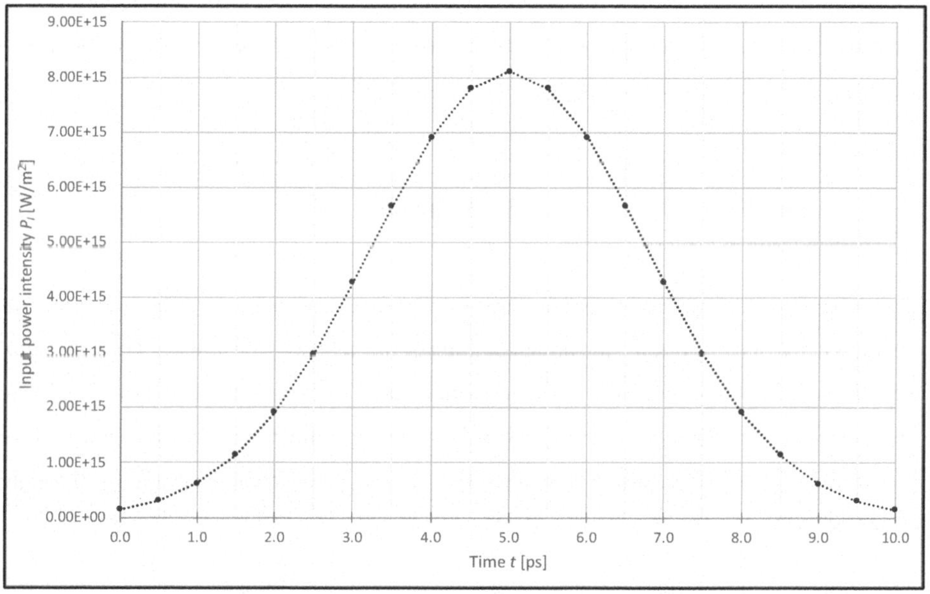

Figure 63. Time distribution of input power intensity of a Gaussian picosecond pulse of a pulse energy of $E_0 = 50$ µJ

Thus, the product of the average of the input power intensity P_I per time segments Δt and the pulse length τ_H gives the average input fluence F, as expressed by equation 33. The following table (table 18) presents the average input fluences at the available pulse repetition rates f of the system for this work.

Table 18: Average input fluence of the laser system

Pulse repetition rate f	Pulse energy E_0	Average input fluence F*	
[kHz]	[µJ]	[J/m^2]	[J/cm^2]
400	125	85423	8.54
500	100	68338	6.83
625	80	54670	5.46
800	63	42711	4.27
1000	50	34169	3.41

* Calculation after equation 33

Finally, the array of the elements after the Beer-Lambert law (equation 25) calculates the progression of the heat source (or power density) through depth over time to allow a visualization of the pulse. Based on the available pulse repetition rates of the system, the related pulse energies and their calculated input power intensities (thanks to equation 32), there are five different heat source Gt distributions presented for this work in the following subchapter *5.3.3 results of the adaptation and conclusions*.

5.3.3 Results of the adaptation and conclusions

The Adaptation of the Beer-Lambert law to picosecond pulse laser beam ablation of cemented WC makes possible the understanding to the heat source Gt distribution through the material. The five available pulse repetition rates f of the laser system defined the available pulse energies E_0, and consequently, fluences F for the analysis. Figures 64 to 68 introduce the five resulting visualizations of the picosecond heat source through cemented WC.

Beer-Lambert law power density distribution	Average input fluence F = 8.54 J/cm^2
Ablated section of material	Pulse repetition rate f = 400 kHz
Thermal affected section of material	Pulse length τ_H = 10 ps

Gt [kW µm^{-3}] Time t [ps]

Depth y [µm]

Gt (y,t)	0	0.5	1	1.5	2	2.5	3	3.5	4	4.5	5	5.5	6	6.5	7	7.5	8	8.5	9	9.5	10
-0.010	7.11	15.19	29.99	54.64	91.91	142.71	204.55	270.65	330.57	372.72	387.93	372.72	330.57	270.65	204.55	142.71	91.91	54.64	29.99	15.19	7.11
-0.015	5.69	12.16	24.00	43.74	73.57	114.23	163.74	216.64	264.61	298.35	310.52	298.35	264.61	216.64	163.74	114.23	73.57	43.74	24.00	12.16	5.69
-0.020	4.55	9.73	19.21	35.01	58.89	91.44	131.06	173.41	211.81	238.81	248.56	238.81	211.81	173.41	131.06	91.44	58.89	35.01	19.21	9.73	4.55
-0.025	3.64	7.79	15.38	28.03	47.14	73.19	104.91	138.81	169.54	191.16	198.96	191.16	169.54	138.81	104.91	73.19	47.14	28.03	15.38	7.79	3.64
-0.030	2.92	6.24	12.31	22.43	37.73	58.59	83.98	111.11	135.71	153.02	159.26	153.02	135.71	111.11	83.98	58.59	37.73	22.43	12.31	6.24	2.92
-0.035	2.33	4.99	9.86	17.96	30.20	46.90	67.22	88.94	108.63	122.48	127.48	122.48	108.63	88.94	67.22	46.90	30.20	17.96	9.86	4.99	2.33
-0.040	1.87	4.00	7.89	14.37	24.18	37.54	53.81	71.19	86.96	98.04	102.04	98.04	86.96	71.19	53.81	37.54	24.18	14.37	7.89	4.00	1.87
-0.045	1.50	3.20	6.31	11.51	19.35	30.05	43.07	56.99	69.61	78.48	81.68	78.48	69.61	56.99	43.07	30.05	19.35	11.51	6.31	3.20	1.50
-0.050	1.20	2.56	5.05	9.21	15.49	24.05	34.48	45.62	55.72	62.82	65.38	62.82	55.72	45.62	34.48	24.05	15.49	9.21	5.05	2.56	1.20
-0.055	0.96	2.05	4.05	7.37	12.40	19.25	27.60	36.51	44.60	50.29	52.34	50.29	44.60	36.51	27.60	19.25	12.40	7.37	4.05	2.05	0.96
-0.060	0.77	1.64	3.24	5.90	9.93	15.41	22.09	29.23	35.70	40.25	41.89	40.25	35.70	29.23	22.09	15.41	9.93	5.90	3.24	1.64	0.77
-0.065	0.61	1.31	2.59	4.72	7.95	12.34	17.68	23.40	28.58	32.22	33.53	32.22	28.58	23.40	17.68	12.34	7.95	4.72	2.59	1.31	0.61
-0.070	0.49	1.05	2.08	3.78	6.36	9.87	14.15	18.73	22.87	25.79	26.84	25.79	22.87	18.73	14.15	9.87	6.36	3.78	2.08	1.05	0.49
-0.075	0.39	0.84	1.66	3.03	5.09	7.90	11.33	14.99	18.31	20.64	21.49	20.64	18.31	14.99	11.33	7.90	5.09	3.03	1.66	0.84	0.39
-0.080	0.32	0.67	1.33	2.42	4.07	6.33	9.07	12.00	14.66	16.52	17.20	16.52	14.66	12.00	9.07	6.33	4.07	2.42	1.33	0.67	0.32
-0.085	0.25	0.54	1.06	1.94	3.26	5.06	7.26	9.61	11.73	13.23	13.77	13.23	11.73	9.61	7.26	5.06	3.26	1.94	1.06	0.54	0.25
-0.090	0.20	0.43	0.85	1.55	2.61	4.05	5.81	7.69	9.39	10.59	11.02	10.59	9.39	7.69	5.81	4.05	2.61	1.55	0.85	0.43	0.20
-0.095	0.16	0.35	0.68	1.24	2.09	3.25	4.65	6.15	7.52	8.48	8.82	8.48	7.52	6.15	4.65	3.25	2.09	1.24	0.68	0.35	0.16
-0.100	0.13	0.28	0.55	0.99	1.67	2.60	3.72	4.93	6.02	6.78	7.06	6.78	6.02	4.93	3.72	2.60	1.67	0.99	0.55	0.28	0.13
-0.105	0.10	0.22	0.44	0.80	1.34	2.08	2.98	3.94	4.82	5.43	5.65	5.43	4.82	3.94	2.98	2.08	1.34	0.80	0.44	0.22	0.10
-0.110	0.08	0.18	0.35	0.64	1.07	1.66	2.39	3.16	3.86	4.35	4.52	4.35	3.86	3.16	2.39	1.66	1.07	0.64	0.35	0.18	0.08

Figure 64: Beer-Lambert law power density distribution of a 10 ps pulse at 400 kHz

Beer-Lambert law power density distribution	Average input fluence F = 6.83 J/cm^2
Ablated section of material	Pulse repetition rate f = 500 kHz
Thermal affected section of material	Pulse length τ_H = 10 ps

Gt [kW µm^{-3}] Time t [ps]

Depth y [µm]

Gt (y,t)	0	0.5	1	1.5	2	2.5	3	3.5	4	4.5	5	5.5	6	6.5	7	7.5	8	8.5	9	9.5	10
-0.010	5.68	12.15	23.99	43.71	73.53	114.17	163.64	216.52	264.46	298.17	310.34	298.17	264.46	216.52	163.64	114.17	73.53	43.71	23.99	12.15	5.68
-0.015	4.55	9.73	19.20	34.99	58.86	91.39	130.99	173.31	211.69	238.68	248.42	238.68	211.69	173.31	130.99	91.39	58.86	34.99	19.20	9.73	4.55
-0.020	3.64	7.79	15.37	28.01	47.11	73.15	104.85	138.73	169.45	191.05	198.85	191.05	169.45	138.73	104.85	73.15	47.11	28.01	15.37	7.79	3.64
-0.025	2.92	6.23	12.30	22.42	37.71	58.56	83.93	111.05	135.64	152.93	159.17	152.93	135.64	111.05	83.93	58.56	37.71	22.42	12.30	6.23	2.92
-0.030	2.33	4.99	9.85	17.95	30.19	46.87	67.18	88.89	108.57	122.41	127.41	122.41	108.57	88.89	67.18	46.87	30.19	17.95	9.85	4.99	2.33
-0.035	1.87	3.99	7.88	14.37	24.16	37.52	53.78	71.15	86.91	97.99	101.99	97.99	86.91	71.15	53.78	37.52	24.16	14.37	7.88	3.99	1.87
-0.040	1.50	3.20	6.31	11.50	19.34	30.03	43.05	56.96	69.57	78.43	81.64	78.43	69.57	56.96	43.05	30.03	19.34	11.50	6.31	3.20	1.50
-0.045	1.20	2.56	5.05	9.20	15.48	24.04	34.46	45.59	55.68	62.78	65.35	62.78	55.68	45.59	34.46	24.04	15.48	9.20	5.05	2.56	1.20
-0.050	0.96	2.05	4.04	7.37	12.39	19.24	27.58	36.49	44.57	50.26	52.31	50.26	44.57	36.49	27.58	19.24	12.39	7.37	4.04	2.05	0.96
-0.055	0.77	1.64	3.24	5.90	9.92	15.40	22.08	29.21	35.68	40.23	41.87	40.23	35.68	29.21	22.08	15.40	9.92	5.90	3.24	1.64	0.77
-0.060	0.61	1.31	2.59	4.72	7.94	12.33	17.67	23.38	28.56	32.20	33.52	32.20	28.56	23.38	17.67	12.33	7.94	4.72	2.59	1.31	0.61
-0.065	0.49	1.05	2.07	3.78	6.36	9.87	14.15	18.72	22.86	25.78	26.83	25.78	22.86	18.72	14.15	9.87	6.36	3.78	2.07	1.05	0.49
-0.070	0.39	0.84	1.66	3.02	5.09	7.90	11.32	14.98	18.30	20.63	21.47	20.63	18.30	14.98	11.32	7.90	5.09	3.02	1.66	0.84	0.39
-0.075	0.31	0.67	1.33	2.42	4.07	6.32	9.06	11.99	14.65	16.52	17.19	16.52	14.65	11.99	9.06	6.32	4.07	2.42	1.33	0.67	0.31
-0.080	0.25	0.54	1.06	1.94	3.26	5.06	7.26	9.60	11.72	13.22	13.76	13.22	11.72	9.60	7.26	5.06	3.26	1.94	1.06	0.54	0.25
-0.085	0.20	0.43	0.85	1.55	2.61	4.05	5.81	7.68	9.39	10.58	11.01	10.58	9.39	7.68	5.81	4.05	2.61	1.55	0.85	0.43	0.20
-0.090	0.16	0.35	0.68	1.24	2.09	3.24	4.65	6.15	7.51	8.47	8.82	8.47	7.51	6.15	4.65	3.24	2.09	1.24	0.68	0.35	0.16
-0.095	0.13	0.28	0.55	0.99	1.67	2.60	3.72	4.92	6.01	6.78	7.06	6.78	6.01	4.92	3.72	2.60	1.67	0.99	0.55	0.28	0.13
-0.100	0.10	0.22	0.44	0.80	1.34	2.08	2.98	3.94	4.81	5.43	5.65	5.43	4.81	3.94	2.98	2.08	1.34	0.80	0.44	0.22	0.10
-0.105	0.08	0.18	0.35	0.64	1.07	1.66	2.38	3.15	3.85	4.34	4.52	4.34	3.85	3.15	2.38	1.66	1.07	0.64	0.35	0.18	0.08
-0.110	0.07	0.14	0.28	0.51	0.86	1.33	1.91	2.53	3.08	3.48	3.62	3.48	3.08	2.53	1.91	1.33	0.86	0.51	0.28	0.14	0.07

Figure 65: Beer-Lambert law power density distribution of a 10 ps pulse at 500 kHz

Beer-Lambert law power density distribution	Average input fluence F = 5.46 J/cm^2
Ablated section of material	Pulse repetition rate f = 625 kHz
Thermal affected section of material	Pulse length τ_H = 10 ps

Gt [kW µm^{-3}] Time t [ps]

Depth y [µm]

Gt(y,t)	0	0.5	1	1.5	2	2.5	3	3.5	4	4.5	5	5.5	6	6.5	7	7.5	8	8.5	9	9.5	10
-0.010	4.55	9.72	19.19	34.97	58.82	91.33	130.91	173.22	211.57	238.54	248.27	238.54	211.57	173.22	130.91	91.33	58.82	34.97	19.19	9.72	4.55
-0.015	3.64	7.78	15.36	27.99	47.09	73.11	104.79	138.65	169.35	190.94	198.73	190.94	169.35	138.65	104.79	73.11	47.09	27.99	15.36	7.78	3.64
-0.020	2.91	6.23	12.30	22.41	37.69	58.52	83.88	110.99	135.56	152.84	159.08	152.84	135.56	110.99	83.88	58.52	37.69	22.41	12.30	6.23	2.91
-0.025	2.33	4.99	9.84	17.94	30.17	46.84	67.14	88.84	108.51	122.34	127.34	122.34	108.51	88.84	67.14	46.84	30.17	17.94	9.84	4.99	2.33
-0.030	1.87	3.99	7.88	14.36	24.15	37.50	53.75	71.11	86.86	97.93	101.93	97.93	86.86	71.11	53.75	37.50	24.15	14.36	7.88	3.99	1.87
-0.035	1.49	3.20	6.31	11.49	19.33	30.01	43.02	56.92	69.53	78.39	81.59	78.39	69.53	56.92	43.02	30.01	19.33	11.49	6.31	3.20	1.49
-0.040	1.20	2.56	5.05	9.20	15.47	24.03	34.44	45.56	55.65	62.75	65.31	62.75	55.65	45.56	34.44	24.03	15.47	9.20	5.05	2.56	1.20
-0.045	0.96	2.05	4.04	7.36	12.39	19.23	27.57	36.47	44.55	50.23	52.28	50.23	44.55	36.47	27.57	19.23	12.39	7.36	4.04	2.05	0.96
-0.050	0.77	1.64	3.23	5.89	9.91	15.39	22.06	29.19	35.66	40.20	41.85	40.20	35.66	29.19	22.06	15.39	9.91	5.89	3.23	1.64	0.77
-0.055	0.61	1.31	2.59	4.72	7.94	12.32	17.66	23.37	28.54	32.18	33.50	32.18	28.54	23.37	17.66	12.32	7.94	4.72	2.59	1.31	0.61
-0.060	0.49	1.05	2.07	3.78	6.35	9.86	14.14	18.71	22.85	25.76	26.81	25.76	22.85	18.71	14.14	9.86	6.35	3.78	2.07	1.05	0.49
-0.065	0.39	0.84	1.66	3.02	5.08	7.90	11.32	14.97	18.29	20.62	21.46	20.62	18.29	14.97	11.32	7.90	5.08	3.02	1.66	0.84	0.39
-0.070	0.31	0.67	1.33	2.42	4.07	6.32	9.06	11.99	14.64	16.51	17.18	16.51	14.64	11.99	9.06	6.32	4.07	2.42	1.33	0.67	0.31
-0.075	0.25	0.54	1.06	1.94	3.26	5.06	7.25	9.59	11.72	13.21	13.75	13.21	11.72	9.59	7.25	5.06	3.26	1.94	1.06	0.54	0.25
-0.080	0.20	0.43	0.85	1.55	2.61	4.05	5.80	7.68	9.38	10.58	11.01	10.58	9.38	7.68	5.80	4.05	2.61	1.55	0.85	0.43	0.20
-0.085	0.16	0.35	0.68	1.24	2.09	3.24	4.65	6.15	7.51	8.47	8.81	8.47	7.51	6.15	4.65	3.24	2.09	1.24	0.68	0.35	0.16
-0.090	0.13	0.28	0.55	0.99	1.67	2.59	3.72	4.92	6.01	6.78	7.05	6.78	6.01	4.92	3.72	2.59	1.67	0.99	0.55	0.28	0.13
-0.095	0.10	0.22	0.44	0.80	1.34	2.08	2.98	3.94	4.81	5.42	5.65	5.42	4.81	3.94	2.98	2.08	1.34	0.80	0.44	0.22	0.10
-0.100	0.08	0.18	0.35	0.64	1.07	1.66	2.38	3.15	3.85	4.34	4.52	4.34	3.85	3.15	2.38	1.66	1.07	0.64	0.35	0.18	0.08
-0.105	0.07	0.14	0.28	0.51	0.86	1.33	1.91	2.52	3.08	3.48	3.62	3.48	3.08	2.52	1.91	1.33	0.86	0.51	0.28	0.14	0.07
-0.110	0.05	0.11	0.22	0.41	0.69	1.07	1.53	2.02	2.47	2.78	2.90	2.78	2.47	2.02	1.53	1.07	0.69	0.41	0.22	0.11	0.05

Figure 66: Beer-Lambert law power density distribution of a 10 ps pulse at 625 kHz

Beer-Lambert law power density distribution	Average input fluence F = 4.27 J/cm^2
Ablated section of material	Pulse repetition rate f = 800 kHz
Thermal affected section of material	Pulse length τ_H = 10 ps

Gt [kW µm^{-3}] Time t [ps]

Depth y [µm]

Gt(y,t)	0	0.5	1	1.5	2	2.5	3	3.5	4	4.5	5	5.5	6	6.5	7	7.5	8	8.5	9	9.5	10
-0.010	3.55	7.60	14.99	27.32	45.96	71.36	102.28	135.32	165.29	186.36	193.96	186.36	165.29	135.32	102.28	71.36	45.96	27.32	14.99	7.60	3.55
-0.015	2.84	6.08	12.00	21.87	36.79	57.12	81.87	108.32	132.30	149.17	155.26	149.17	132.30	108.32	81.87	57.12	36.79	21.87	12.00	6.08	2.84
-0.020	2.28	4.87	9.61	17.51	29.45	45.72	65.53	86.71	105.90	119.41	124.28	119.41	105.90	86.71	65.53	45.72	29.45	17.51	9.61	4.87	2.28
-0.025	1.82	3.90	7.69	14.01	23.57	36.60	52.46	69.41	84.77	95.58	99.48	95.58	84.77	69.41	52.46	36.60	23.57	14.01	7.69	3.90	1.82
-0.030	1.46	3.12	6.16	11.22	18.87	29.29	41.99	55.56	67.86	76.51	79.63	76.51	67.86	55.56	41.99	29.29	18.87	11.22	6.16	3.12	1.46
-0.035	1.17	2.50	4.93	8.98	15.10	23.45	33.61	44.47	54.32	61.24	63.74	61.24	54.32	44.47	33.61	23.45	15.10	8.98	4.93	2.50	1.17
-0.040	0.93	2.00	3.94	7.19	12.09	18.77	26.90	35.60	43.48	49.02	51.02	49.02	43.48	35.60	26.90	18.77	12.09	7.19	3.94	2.00	0.93
-0.045	0.75	1.60	3.16	5.75	9.68	15.02	21.54	28.49	34.80	39.24	40.84	39.24	34.80	28.49	21.54	15.02	9.68	5.75	3.16	1.60	0.75
-0.050	0.60	1.28	2.53	4.60	7.75	12.03	17.24	22.81	27.86	31.41	32.69	31.41	27.86	22.81	17.24	12.03	7.75	4.60	2.53	1.28	0.60
-0.055	0.48	1.02	2.02	3.69	6.20	9.63	13.80	18.26	22.30	25.14	26.17	25.14	22.30	18.26	13.80	9.63	6.20	3.69	2.02	1.02	0.48
-0.060	0.38	0.82	1.62	2.95	4.96	7.71	11.05	14.61	17.85	20.13	20.95	20.13	17.85	14.61	11.05	7.71	4.96	2.95	1.62	0.82	0.38
-0.065	0.31	0.66	1.30	2.36	3.97	6.17	8.84	11.70	14.29	16.11	16.77	16.11	14.29	11.70	8.84	6.17	3.97	2.36	1.30	0.66	0.31
-0.070	0.25	0.53	1.04	1.89	3.18	4.94	7.08	9.36	11.44	12.90	13.42	12.90	11.44	9.36	7.08	4.94	3.18	1.89	1.04	0.53	0.25
-0.075	0.20	0.42	0.83	1.51	2.55	3.95	5.66	7.50	9.15	10.32	10.74	10.32	9.15	7.50	5.66	3.95	2.55	1.51	0.83	0.42	0.20
-0.080	0.16	0.34	0.66	1.21	2.04	3.16	4.53	6.00	7.33	8.26	8.60	8.26	7.33	6.00	4.53	3.16	2.04	1.21	0.66	0.34	0.16
-0.085	0.13	0.27	0.53	0.97	1.63	2.53	3.63	4.80	5.87	6.61	6.88	6.61	5.87	4.80	3.63	2.53	1.63	0.97	0.53	0.27	0.13
-0.090	0.10	0.22	0.43	0.78	1.31	2.03	2.91	3.84	4.70	5.29	5.51	5.29	4.70	3.84	2.91	2.03	1.31	0.78	0.43	0.22	0.10
-0.095	0.08	0.17	0.34	0.62	1.04	1.62	2.33	3.08	3.76	4.24	4.41	4.24	3.76	3.08	2.33	1.62	1.04	0.62	0.34	0.17	0.08
-0.100	0.06	0.14	0.27	0.50	0.84	1.30	1.86	2.46	3.01	3.39	3.53	3.39	3.01	2.46	1.86	1.30	0.84	0.50	0.27	0.14	0.06
-0.105	0.05	0.11	0.22	0.40	0.67	1.04	1.49	1.97	2.41	2.72	2.83	2.72	2.41	1.97	1.49	1.04	0.67	0.40	0.22	0.11	0.05
-0.110	0.04	0.09	0.17	0.32	0.54	0.83	1.19	1.58	1.93	2.17	2.26	2.17	1.93	1.58	1.19	0.83	0.54	0.32	0.17	0.09	0.04

Figure 67: Beer-Lambert law power density distribution of a 10 ps pulse at 800 kHz

Beer-Lambert law power density distribution	Average input fluence F = 3.41 J/cm²
Ablated section of material	Pulse repetition rate f = 1000 kHz
Thermal affected section of material	Pulse length τ_H = 10 ps

Gt [kW µm⁻³] Time t [ps]

Depth y [µm]

Gt [y,t]	0	0.5	1	1.5	2	2.5	3	3.5	4	4.5	5	5.5	6	6.5	7	7.5	8	8.5	9	9.5	10
-0.010	2.84	6.08	12.00	21.86	36.76	57.08	81.82	108.26	132.23	149.09	155.17	149.09	132.23	108.26	81.82	57.08	36.76	21.86	12.00	6.08	2.84
-0.015	2.27	4.86	9.60	17.50	29.43	45.69	65.49	86.66	105.84	119.34	124.21	119.34	105.84	86.66	65.49	45.69	29.43	17.50	9.60	4.86	2.27
-0.020	1.82	3.89	7.69	14.00	23.56	36.58	52.43	69.37	84.72	95.53	99.42	95.53	84.72	69.37	52.43	36.58	23.56	14.00	7.69	3.89	1.82
-0.025	1.46	3.12	6.15	11.21	18.86	29.28	41.96	55.52	67.82	76.46	79.59	76.46	67.82	55.52	41.96	29.28	18.86	11.21	6.15	3.12	1.46
-0.030	1.17	2.49	4.92	8.97	15.09	23.44	33.59	44.45	54.29	61.21	63.70	61.21	54.29	44.45	33.59	23.44	15.09	8.97	4.92	2.49	1.17
-0.035	0.93	2.00	3.94	7.18	12.08	18.76	26.89	35.58	43.45	48.99	50.99	48.99	43.45	35.58	26.89	18.76	12.08	7.18	3.94	2.00	0.93
-0.040	0.75	1.60	3.16	5.75	9.67	15.02	21.52	28.48	34.78	39.22	40.82	39.22	34.78	28.48	21.52	15.02	9.67	5.75	3.16	1.60	0.75
-0.045	0.60	1.28	2.53	4.60	7.74	12.02	17.23	22.80	27.84	31.39	32.67	31.39	27.84	22.80	17.23	12.02	7.74	4.60	2.53	1.28	0.60
-0.050	0.48	1.02	2.02	3.68	6.20	9.62	13.79	18.25	22.29	25.13	26.15	25.13	22.29	18.25	13.79	9.62	6.20	3.68	2.02	1.02	0.48
-0.055	0.38	0.82	1.62	2.95	4.96	7.70	11.04	14.61	17.84	20.11	20.93	20.11	17.84	14.61	11.04	7.70	4.96	2.95	1.62	0.82	0.38
-0.060	0.31	0.66	1.30	2.36	3.97	6.16	8.84	11.69	14.28	16.10	16.76	16.10	14.28	11.69	8.84	6.16	3.97	2.36	1.30	0.66	0.31
-0.065	0.25	0.53	1.04	1.89	3.18	4.93	7.07	9.36	11.43	12.89	13.41	12.89	11.43	9.36	7.07	4.93	3.18	1.89	1.04	0.53	0.25
-0.070	0.20	0.42	0.83	1.51	2.54	3.95	5.66	7.49	9.15	10.32	10.74	10.32	9.15	7.49	5.66	3.95	2.54	1.51	0.83	0.42	0.20
-0.075	0.16	0.34	0.66	1.21	2.04	3.16	4.53	6.00	7.32	8.26	8.59	8.26	7.32	6.00	4.53	3.16	2.04	1.21	0.66	0.34	0.16
-0.080	0.13	0.27	0.53	0.97	1.63	2.53	3.63	4.80	5.86	6.61	6.88	6.61	5.86	4.80	3.63	2.53	1.63	0.97	0.53	0.27	0.13
-0.085	0.10	0.22	0.43	0.78	1.30	2.03	2.90	3.84	4.69	5.29	5.51	5.29	4.69	3.84	2.90	2.03	1.30	0.78	0.43	0.22	0.10
-0.090	0.08	0.17	0.34	0.62	1.04	1.62	2.32	3.08	3.76	4.24	4.41	4.24	3.76	3.08	2.32	1.62	1.04	0.62	0.34	0.17	0.08
-0.095	0.06	0.14	0.27	0.50	0.84	1.30	1.86	2.46	3.01	3.39	3.53	3.39	3.01	2.46	1.86	1.30	0.84	0.50	0.27	0.14	0.06
-0.100	0.05	0.11	0.22	0.40	0.67	1.04	1.49	1.97	2.41	2.71	2.82	2.71	2.41	1.97	1.49	1.04	0.67	0.40	0.22	0.11	0.05
-0.105	0.04	0.09	0.17	0.32	0.54	0.83	1.19	1.58	1.93	2.17	2.26	2.17	1.93	1.58	1.19	0.83	0.54	0.32	0.17	0.09	0.04
-0.110	0.03	0.07	0.14	0.25	0.43	0.67	0.95	1.26	1.54	1.74	1.81	1.74	1.54	1.26	0.95	0.67	0.43	0.25	0.14	0.07	0.03

Figure 68: Beer-Lambert law power density distribution of a 10 ps pulse at 1000 kHz

The orange section in the visualizations represents the ablated region of the material; it illustrates the power density that delivers the minimum energy to equal or exceed the threshold fluence of the material, and thus ablate the material by equaling or exceeding its ablation threshold. The other colored sections in the visualizations correspond to heated material which does not reach the ablation threshold.

Figure 69 summarizes the relationship of average input fluences and their respective resulting depths in the analysis for all pulse repetition rates. Among the pulse repetitions rates, it can be observed a constant relationship to their respective average input fluences, i.e., pulse energy.

The distribution of the energy into the material required a higher density of it in the surface until it vanishes at the analyzed depth of the material after the Beer-Lambert law. For example, the amount of average input power density *Gt* irradiated at a *y* depth of the material can be observed in figure 70 at a pulse repetition depth of $f = 1000$ kHz and for a laser average power of $P_L = 50$ W. This fact supports the understanding of a pulse irradiating a material in terms of fluence and power density.

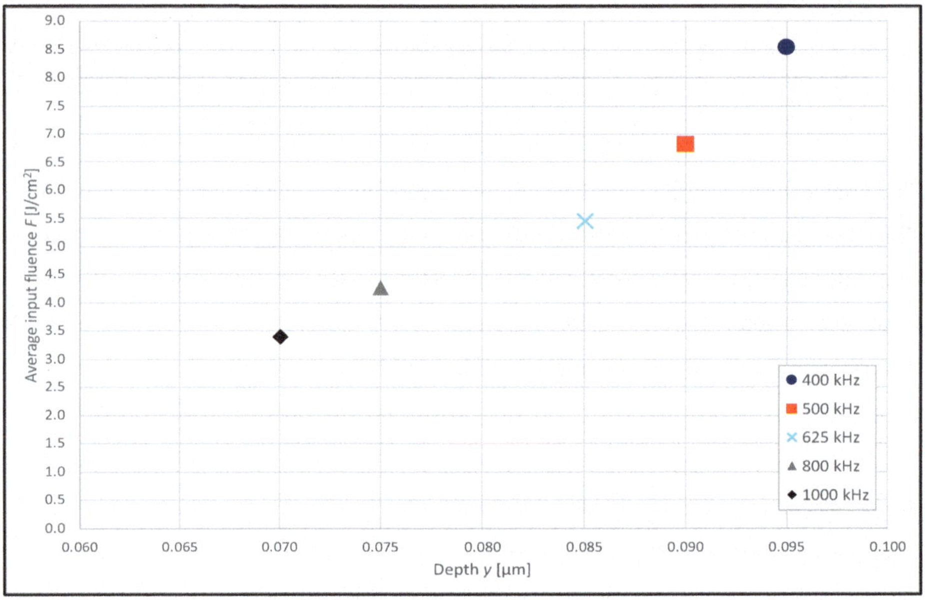

Figure 69: Average input fluence F vs ablation depth y of a 10 ps pulse at $f = 400$ kHz, 500 kHz, 625 kHz, 800 kHz and 1000 kHz after the Beer-Lambert law analysis

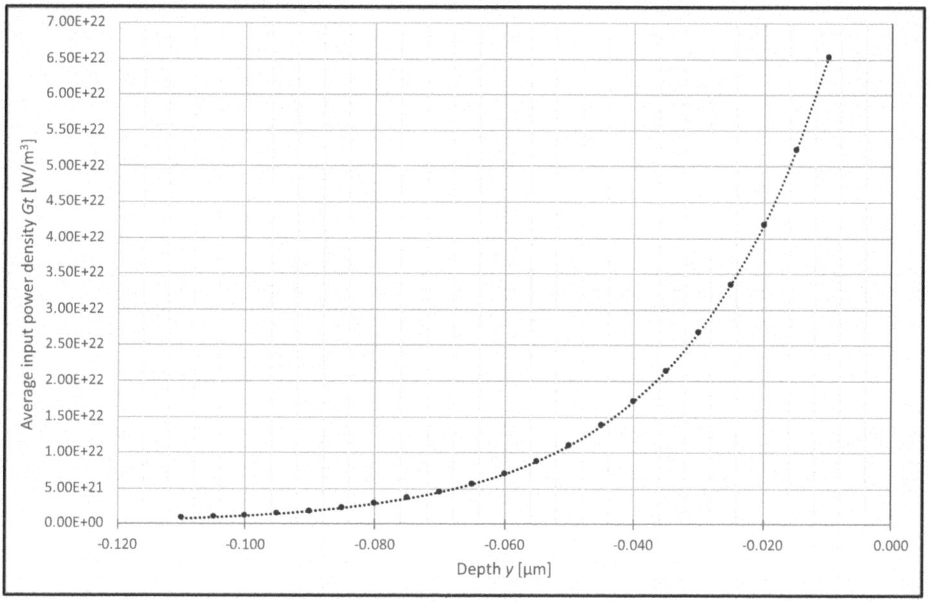

Figure 70: Distribution of the average input power density *Gt* of a 10 ps pulse into the material (depth *y*) at a pulse repetition rate of *f* = 1000 kHz

In conclusion, different pulse energies were analyzed under the theory of the Beer-Lambert law. An analysis resulted in the visualization of a pulse at different pulse repetition rates and considering aspects, such as, thermal material properties, Gaussian behavior and reflectivity, among others. Important for its validity is their comparison to experimental results. Therefore, figure 71 contrasts both values: results coming from the Beer-Lambert law analysis in this chapter and validation experiments. The execution of the experiments was done in accordance with the laser system characteristics; the pulses were isolated to its minimum overlap to obtain single pulses (*PO* equation 15), e.g., *PO* = 11% and 29%. The measurement of the depth was target to the section where no overlap existed. A slight overlap of *PO* = 55% to 64% could be possible for some cases and neglected; values above this range were avoided. The experimental results are available for the pulse repetition rates of *f* = 400 kHz, 500 kHz, 800 kHz and 1000 kHz.

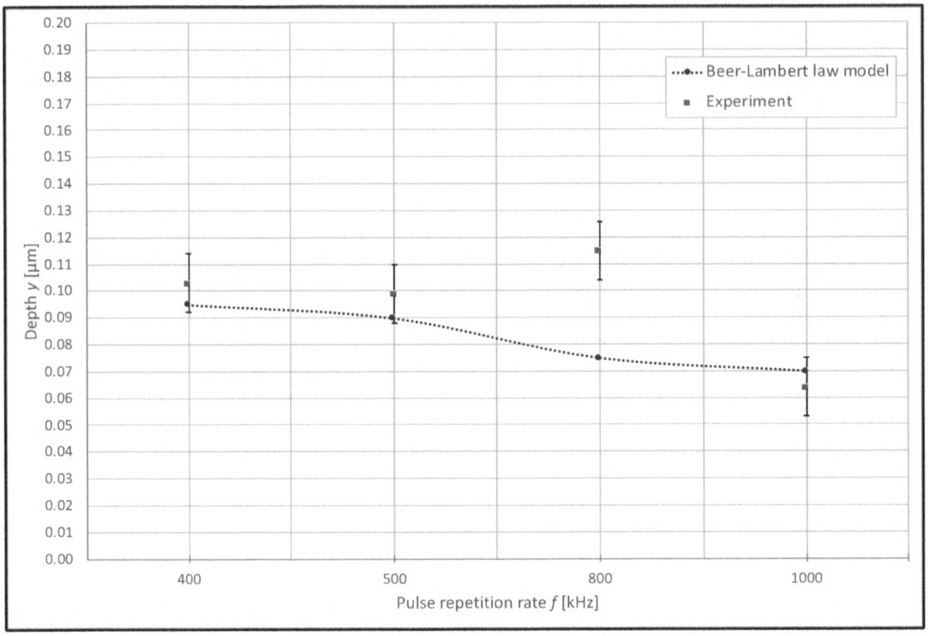

Figure 71: Experimental validation of the adaptation of the Beer-Lambert law to picosecond pulse laser ablation

The experimental results present a similar acceptable behavior when compared to the results of the analytical approach. A higher contrast of the results is shown for $f = 800$ kHz. Further experimental analysis and validation under controlled conditions, e.g., isolated pulses, can be done in this subject.

5.4 Development of the multi-pulse laser ablation model

The understanding of the behavior of one pulse irradiating the material in 5.3 is basic for the understanding of multi-pulse irradiation. The emission of multiple pulses is a fundamental operation in laser material processing and its efficient and adequate application determines the results in terms of quality and productivity, i.e., quality, cost and time. For this reason, a detailed analysis is required to identify the optimal parameter combination for ablation based on the Beer-Lambert law knowledge from one pulse and the selection of involved settings and parameters that command the delivery of multiple pulses on the material, such as, pulse repetition rate f, scanner speed v_s and the resulting pulse overlap PO, among others. A development of a multi-pulse ablation model targets the obtention of a controllable description of the ablation model under the system conditions of this work and on cemented WC; this approach can be contrasted by an experimental validation to evaluate its scope. Moreover, this analysis of multi-pulse laser ablation includes two perspectives of multi-pulse irradiation; the first corresponds to a multi-pulse irradiation with single-pulse mode, which is the default mode with no partition of the pulses; and, the second corresponds to the multi-pulse irradiation with burst-mode (partition of a pulse into burst, as explained in subchapter 2.1.5). The following lines describe the analysis of multi-pulse laser ablation starting with a theoretical background; continuing with the development of the multi-pulse laser ablation model with single-pulse mode and its experimental validation; likewise, developing the multi-pulse laser ablation model with burst-mode and its experimental validation; and concluding with final remarks.

5.4.1 Theoretical background of the model

The approach in this part targets the determination of the ablation depth Z_{abl} as a result of modeling laser beam ablation when the system irradiates multiple pulses. For this reason, the concept of Z_{abl} is presented in the following lines. The investigations on the logarithmic relationship of ablation depth and intensity, described by equation 5 early in this work [24; 142], supports the definition of the behavior of ablation depth as a function of optical penetration depth δ, input fluence F and threshold fluence F_{th} [45; 100]. This behavior is expressed in equation 46 and it describes the ablation depth for the irradiation of one single pulse.

$$Z_{abl} = \delta \, ln\left(\frac{F}{F_{th}}\right) \tag{46}$$

Multiple pulses, that are interacting in the ablation, change the behavior of the process due to the incubation phenomenon of pulses irradiating the same spot on the material. The high repetition rates, the diameter of the laser spot and the speed of the scanner system determine the number of pulses hitting the same area on the material even when the scanner is moving the laser spot through a length in a time period. Figure 72 illustrates the idea of pulses being irradiated on the same area and the elements for the determination of the number of pulses *N*.

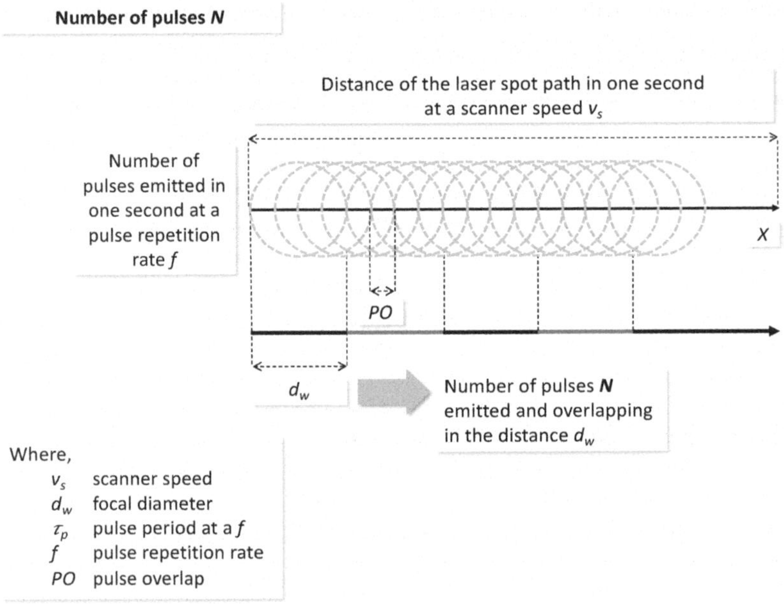

Figure 72: Determination of the number of pulses *N* on a same area

The scanner system positions the laser spot along a distance at a scanner speed v_s and emits a number of pulses according to the selected pulse repetition rate f at an equivalent period of time; the number of pulses *N* hitting the same length, the focal diameter d_w for this case, can be quantified by calculating the number of pulses emitted and overlapping on this d_w distance at a known v_s and f, as expressed in equation 47.

$$N = \frac{f \cdot d_w}{v_s} \qquad (47)$$

The definition of the number of pulses *N* on the same spot together with the application of the incubation factor *S* (discussed in subchapter 5.1) allows the identification of the threshold fluence F_{th} of each subsequent pulse on the same surface spot and the gradual fulfillment of equation 46 after the threshold fluence of a single pulse is defined. Thus, subchapter 5.3.2 developed a methodology for the identification of the threshold fluence of one pulse for cemented tungsten carbide from its material properties and serves in this chapter as the initial threshold fluence $F_{th(1)}$ for multi-pulse irradiation. Therefore, the threshold fluence in equation 9 can be completed (by the elements $F_{th(1)}$, S and N) and the different threshold fluences F_{th} for different numbers of pulses can be calculated. The definition of ablation depth Z_{abl}, based on equation 46, is adapted to N and F_{th} for multiple pulses and is expressed in the following logarithmic equation 48.

$$Z_{abl} = N \cdot \delta \, ln\left(\frac{F}{F_{th}}\right) \quad (48)$$

The second perspective to ablate the material and important in this work to be included in a model corresponds to the application of the burst-mode.

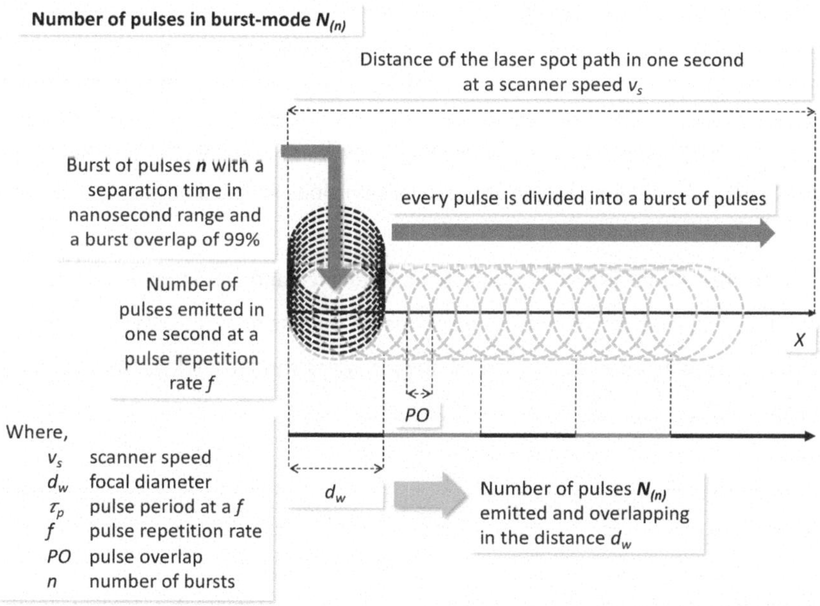

Figure 73: Determination of the number of pulses in burst-mode $N_{(n)}$ on a same area

The possibility of the laser system to use the partition of pulses or burst-mode implies modifications in the calculation of the elements in the ablation process already described for multi-pulse ablation with single-pulse mode. First, the calculation of the effective number of pulses hitting the material must include the separation of a pulse into burst. The burst of a pulse is separated in a nanosecond time range followed by another burst of the consecutive pulses; the overlap of the bursts among them is considered to be of 99% due to their ultra-short separation time. The division of these groups of partitioned pulses or bursts, as described in chapter 2, is commanded by the pulse repetition rate in form of the corresponding pulse period τ_p. Figure 73 illustrates the irradiation of pulses in burst-mode and describes the elements to determine the number of pulses $N_{(n)}$.

Equation 49 modifies the expression of equation 48 to include the partition of the pulses into bursts. The element n corresponds to the number of bursts selected in the laser system burst-mode. The burst-mode can be set from 1 to 10 (integers) in the laser system operated in this work and the selection of $n = 1$ is understood as the absence of bursts.

$$N_{(n)} = \frac{f \cdot d_w}{v_s} \cdot n \qquad (49)$$

Basically, the number of pulses in burst-mode can be expressed as the product of the number of pulses N and the number of bursts or partitions selected n.

Threshold fluence for the pulses in burst-mode is adapted and expressed in equation 50. It describes the total number of pulses in groups of bursts irradiating the material while considering their incubation effect. The element $F_{th(1)}$ is similarly taken from the analysis in subchapter 5.3.2.

$$F_{th} = F_{th(1)} \cdot N_{(n)}^{S-1} \qquad (50)$$

The determination of the ablation threshold F_{th} allows to calculate the ablation depth Z_{abl} for the total number of pulses including bursts of pulses $N_{(n)}$. Based on equation 48, the ablation depth equation modifies the number of pulses N in burst-mode $N_{(n)}$, as seen in equation 51.

$$Z_{abl} = N_{(n)} \cdot \delta \, ln\left(\frac{F_{(n)}}{F_{th}}\right) \tag{51}$$

An important factor to define in the determination of the ablation depth in equation 51 is the correct calculation of input fluence F for burst-mode. The amount of fluence of a pulse has to be adapted to the pulse energy E_0 of one single burst; and consequently, the input fluence F has to be adapted to a single burst. Thus, the input fluence F of a pulse has to be divided by the number of bursts or partitions n to determine a correct single burst value of $F_{(n)}$. Based on the laser source information [129], the partition of a pulse into bursts of pulses is uniform in this work, i.e., the resulting burst fluences or energies are equally distributed for all bursts. In contrast, other studies had shown a different energy level of the bursts [143], e.g., an exponential behavior of subsequent bursts by having a reduction in burst energy through time. This aspect of burst behavior and generation is not discussed or analyzed in this work but is worthy of further studies.

In the following lines, the theoretical knowledge is implemented to execute an ablation model for multi-pulse laser ablation with single-pulse mode and burst-mode for the picosecond laser beam ablation of cemented WC. After each mode analysis, an experimental validation of the models is presented.

5.4.2 Multi-pulse laser ablation model with single-pulse mode

The model of multi-pulse laser ablation irradiating single pulses (absence of burst-mode $n > 1$) is based on the calculation of the number of pulses N that are irradiated on the same area, the determination of the threshold fluence F_{th} for the number of pulses and the calculation of the ablation depth Z_{abl} for these conditions. As presented in the theoretical background, equation 47 for N, equation 9 for F_{th} and equation 48 for Z_{abl} guide this analysis, respectively. Moreover, subchapter 5.1, 5.2 and 5.3 offer the elements to complete this approach. For example, the determination of the incubation factor S, identification of the optical penetration depth δ of cemented WC, and the calculation of the basic threshold fluence $F_{th(1)}$ of cemented WC based on the heat transfer equations are some of the important supporting elements for the analysis. The objective of the model is to describe the behavior of picosecond laser beam ablation on cemented WC in terms of ablation depth Z_{abl}. Table 19 presents a summary of the properties and parameters of the system that are used in the execution of the model for multi-pulse ablation; similarly, their respective values and settings are listed.

Table 19: Summary of parameters and settings used in the model

Properties / parameters	Element	Value	Unit	Source
Laser beam power P_L	Laser system	100, 50	[%][W]	Measured
Beam radius ω_0	Laser system	14	[µm]	Measured
Pulse length τ_H	Laser system	10	[ps]	Data sheet
Pulse and burst mode BM	Laser system	1 (no bursts)	[-]	Data sheet
Pulse repetition rate f	Laser system	400, 500, 800, 1000	[kHz]	Data sheet
Scanner speed v_s	Scanner system	500, 1000, 2000, 5000	[mm/s]	Data sheet
Incubation factor S	Material	0.75	[-]	Subchapter 5.1
Reflectivity R	Material	0.33	[-]	Subchapter 5.2
Optical penetration depth δ	Material	2.25 x 10^{-8}	[m]	Subchapter 5.3
Threshold fluence $F_{th(1)}$	Material	0.1964	[J/cm^2]	Subchapter 5.3

Threshold fluence $F_{th(1)}$ is taken from the analysis of the Beer-Lambert law and heat transfer presented in subchapter 5.3

In order to identify the threshold fluence F_{th} for each set of number of pulses N in equation 9, it is necessary to determine N. Based on equation 47, table 20 presents the interactions of pulse repetition rate and scanner speed to determine N and its resulting pulse overlap PO under the selected settings of the analysis in this work.

Table 20: Interactions of pulse repetition rate, scanner speed, pulse overlap and number of pulses on same area

Pulse repetition rate	Scanner speed	Pulse overlap ♥♦	Number of pulses ♥
f	v_S	PO	N
[kHz]	[mm/s]	[-]	[-]
400	500	0.96	22.4
	1000	0.91	11.2
	2000	0.82	5.6
	5000	0.55	2.24
500	500	0.96	28
	1000	0.93	14
	2000	0.86	7
	5000	0.64	2.8
800	500	0.98	44.8
	1000	0.96	22.4
	2000	0.91	11.2
	5000	0.78	4.48
1000	500	0.98	56
	1000	0.96	28
	2000	0.93	14
	5000	0.82	5.6

♥ Calculations for a beam radius ω_0 of 14 µm
♦ Calculations for a pulse length τ_H of 10 ps

Once N is calculated, the corresponding threshold fluence for each number of pulses is determined with the application of equation 9 as mentioned before. Incubation factor S and the threshold fluence for one pulse $F_{th(1)}$ complement this calculation. The following table 21 lists the resulting values of F_{th}.

Table 21: Modelled threshold fluence F_{th} for N

Pulse repetition rate	Scanner speed	Number of pulses	Threshold fluence *	
f	v_S	N	F_{th}	
[kHz]	[mm/s]	[-]	[J/m²]	[J/cm²]
400	500	22.4	903	0.09
	1000	11.2	1073	0.107
	2000	5.6	1277	0.12
	5000	2.24	1605	0.16
500	500	28	854	0.08
	1000	14	1015	0.101
	2000	7	1207	0.12
	5000	2.8	1518	0.15
800	500	44.8	759	0.07
	1000	22.4	903	0.09
	2000	11.2	1073	0.107
	5000	4.48	1350	0.13
1000	500	56	718	0.07
	1000	28	854	0.08
	2000	14	1015	0.101
	5000	5.6	1277	0.12

* Based on the threshold fluence for one pulse $F_{th(1)} = 0.1964$ J/cm²

The determination of F_{th} sets the border of the energy which is required to allow ablation. This required energy, emitted from the laser, has to be calculated taking into consideration two aspects. The first, the Gaussian distribution and Beer-Lambert law analysis already presented earlier in this work (resulted from equations 32 and 33 in subchapter 5.3); and second, the laser and scanner conditions of the system, e.g., pulse repetition rate, laser beam power, beam radius, etc. These considerations allowed the determination of the input fluence F to target the modeling of the ablation depth on cemented WC. Table 18 in subchapter 5.3 presented the average input fluence F, for the four pulse repetition rates f and for a laser beam power of P_L = 100%, that are used for the model of this work. A summarized version of the average input fluence table 18 is presented in table 22.

Finally, the modeling of the ablation depth Z_{abl} in cemented WC contains all required elements for the execution of equation 48. This calculation employs four elements: the optical penetration depth δ of the picosecond beam at λ = 1064 nm in cemented WC, the threshold fluence F_{th} for the different set of N's, the average input fluence F irradiated from the source and the number of pulses N.

Table 22: Average input fluence of the laser system (summarized version)

Pulse repetition rate f	Average input fluence F
[kHz]	[J/m²]
400	85423
500	68338
625	54670
800	42711
1000	34169

Beam power P_L of 50 W (100%)
Beam radius ω_0 of 14 µm

The results of the model are illustrated in the following figures; they present the behavior of the threshold fluence F_{th} and ablation depth Z_{abl} at a pulse repetition rate f versus different levels of scanner speed v_s. Figure 74 presents on the left the results of the modelled threshold fluence F_{th} versus scanner speed v_s at a pulse repetition rate $f = 400$ kHz, and contrasting on the right, the modelled ablation depth Z_{abl} versus scanner speed v_s at the same pulse repetition rate f. Figures 75 to 77 illustrate likewise the modelled outcome at pulse repetition rates $f = 500$ kHz, 800 kHz and 1000 kHz, respectively.

An important aspect in the understanding of the figures is the fact that lower scanner speeds at a determined pulse repetition rate f imply a higher number of pulses irradiated on the material and vice versa.

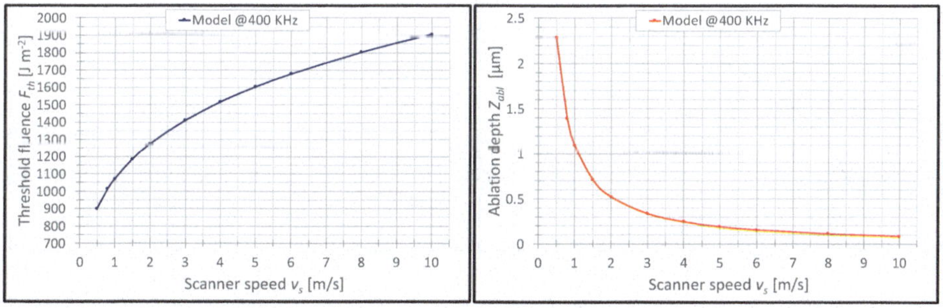

Figure 74: Results of the modeling of threshold fluence F_{th} (left) and ablation depth Z_{abl} (right) at $f = 400$ kHz vs scanner speed v_s

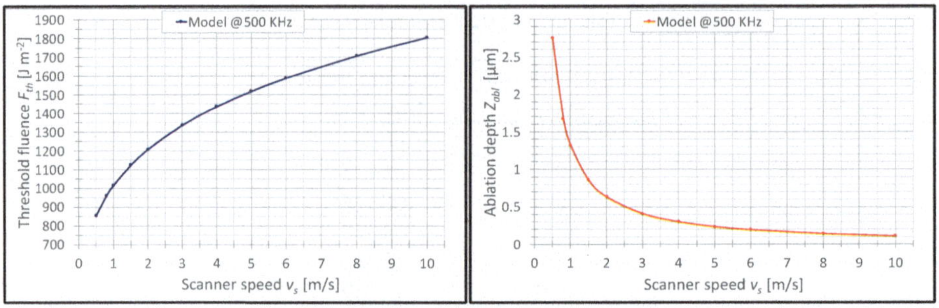

Figure 75: Results of the modeling of threshold fluence F_{th} (left) and ablation depth Z_{abl} (right) at f = 500 kHz vs scanner speed v_s

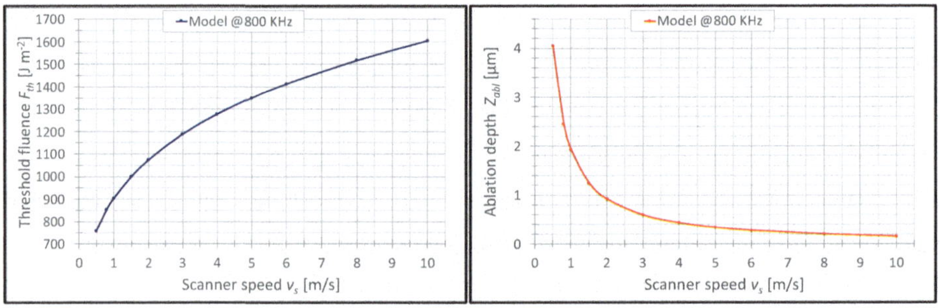

Figure 76: Results of the modeling of threshold fluence F_{th} (left) and ablation depth Z_{abl} (right) at f = 800 kHz vs scanner speed v_s

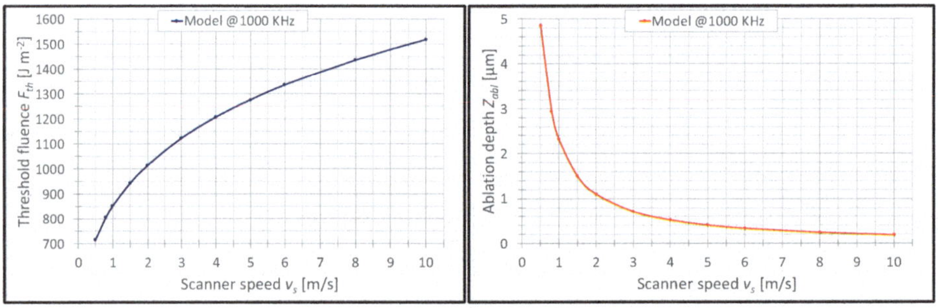

Figure 77: Results of the modeling of threshold fluence F_{th} (left) and ablation depth Z_{abl} (right) at f = 1000 kHz vs scanner speed v_s

Therefore, and based on the incubation effect theory, a high number of pulses and energy irradiation reduces the ablation threshold of the material and higher ablation depth can be observed. On the other side, higher scanner speeds at a determined pulse repetition rate represent a lower number of pulses and causes an increase of the ablation threshold and thus a lower ablation.

Even when the increase of the pulse repetition rate *f* means a decrease of the pulse energy E_0, a higher ablation can be observed thanks to the high number of pulses *N* on the same area and its influence reducing threshold fluence F_{th}.

5.4.3. Experimental validation of the multi-pulse model with single-pulse mode

In order to take advantage of the ablation model for multi-pulse ablation, it is necessary to experimentally validate the model results. Thus, experiments were done maintaining the same parameters, settings and conditions used in the model.

Based on the experimental set-up of this work (system equipment and material in subchapter 4.2), lines of 1 mm length were programmed to be irradiated on a prepared plate of cemented WC. Every line was irradiated unidirectionally at a specific and desired set of parameters and settings. The main controlled parameters, according to the model, pulse repetition rate *f*, scanner speed v_s and burst-mode *BM*. The interaction of *f* and v_s defined a specific pulse overlap *PO* and number of pulses *N*. The following parameters were held constant: laser beam power P_L (100%), focus position of a z_f = 0 mm and incidence angle γ at 90°. After irradiation, the experiments were measured in the confocal microscope (as described in subchapter 4.2.4) using an average range of 200 measurement lines and targeting their ablated cavity (ablation depth Z_{abl}).

After the experiments are executed and measured, the results are listed in the following table 23. The parameters, settings and modeled results are also included for comparison and better understanding.

Table 23: Results of the experimentation

Pulse repetition rate	Scanner speed	Number of pulses	Ablation depth	
f	v_s	N	Model Z_{abl}	**Experiment Z_{abl}**
[kHz]	[mm/s]	[-]	[μm]	[μm]
400	500	22.4	2.29	**1.62**
	1000	11.2	1.10	**1.17**
	2000	5.6	0.53	**0.96**
	5000	2.24	0.20	**0.92**
500	500	28	2.76	**2.01**
	1000	14	1.32	**1.30**
	2000	7	0.63	**0.96**
	5000	2.8	0.24	**1.03**
800	500	44.8	4.06	**2.75**
	1000	22.4	1.94	**1.58**
	2000	11.2	0.93	**1.19**
	5000	4.48	0.35	**1.11**
1000	500	56	4.86	**3.28**
	1000	28	2.32	**2.12**
	2000	14	1.11	**1.20**
	5000	5.6	0.41	**0.76**

Finally, the experimental results of the laser beam ablation can be contrasted to the model to evaluate its validity. The following figures (figures 78 to 81) illustrate the behaviors of both model and experimental ablations in terms of ablation depth Z_{abl} versus scanner speed v_s and at the four selected pulse repetition rates f of this work.

It can be observed that the experimental results follow a similar behavior when compared to the model results. At lower scanner speed, the model forecasts a higher ablation than the experiments. On the other hand, at faster scanner speeds, the experiment shows higher ablation. Moreover, higher ablation was observed with the irradiation of high number of pulses like in the case of high pulse repetition rates, e.g., $f = 1000$ kHz, at lower scanner speeds. Constant ablation was observed among all pulse repetitions rates at high scanner speeds from 0.8 to 1.2 μm. Figure 82 summarizes all results with single-pulse at all pulse repetition rates for comparison (error is omitted for visualization).

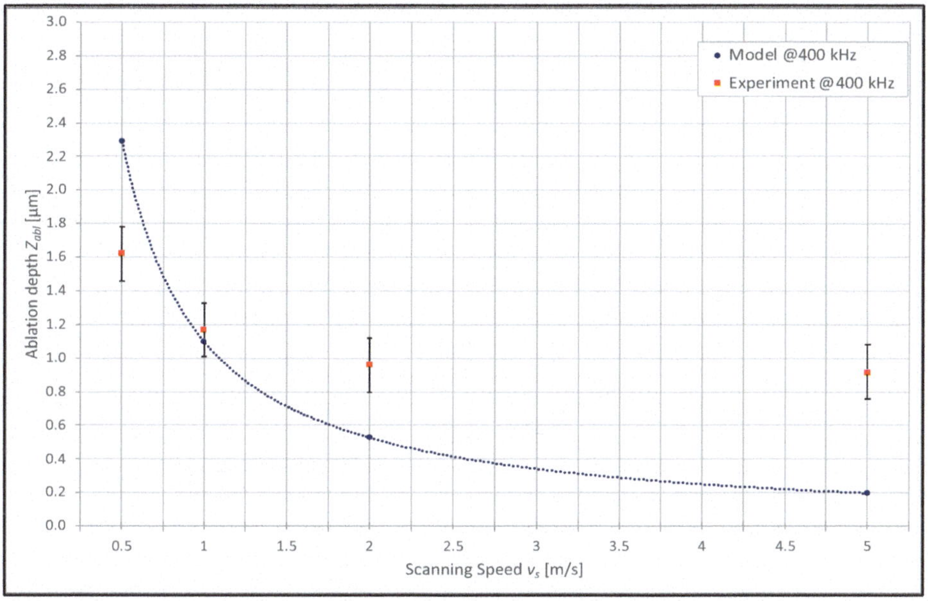

Figure 78: Experimental validation of the multi-pulse picosecond laser beam ablation model with single-pulse on cemented WC at a pulse repetition rate of f = 400 kHz

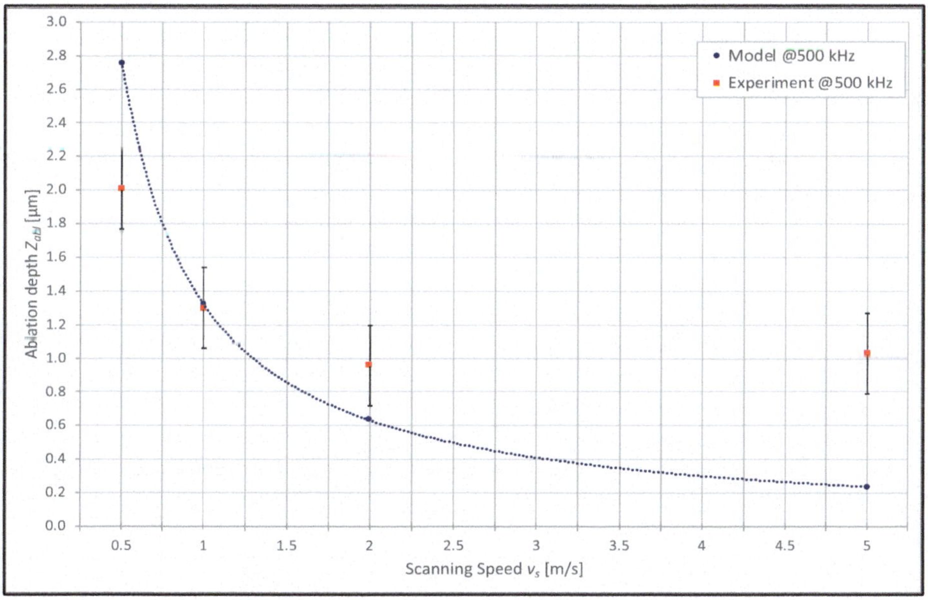

Figure 79: Experimental validation of the multi-pulse picosecond laser beam ablation model with single-pulse on cemented WC at a pulse repetition rate of f = 500 kHz

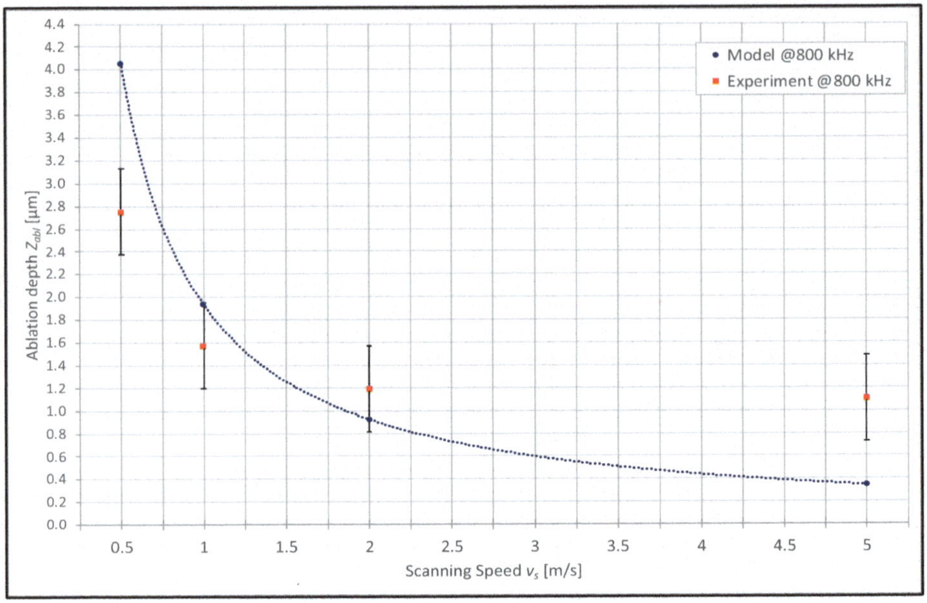

Figure 80: Experimental validation of the multi-pulse picosecond laser beam ablation model with single-pulse on cemented WC at a pulse repetition rate of f = 800 kHz

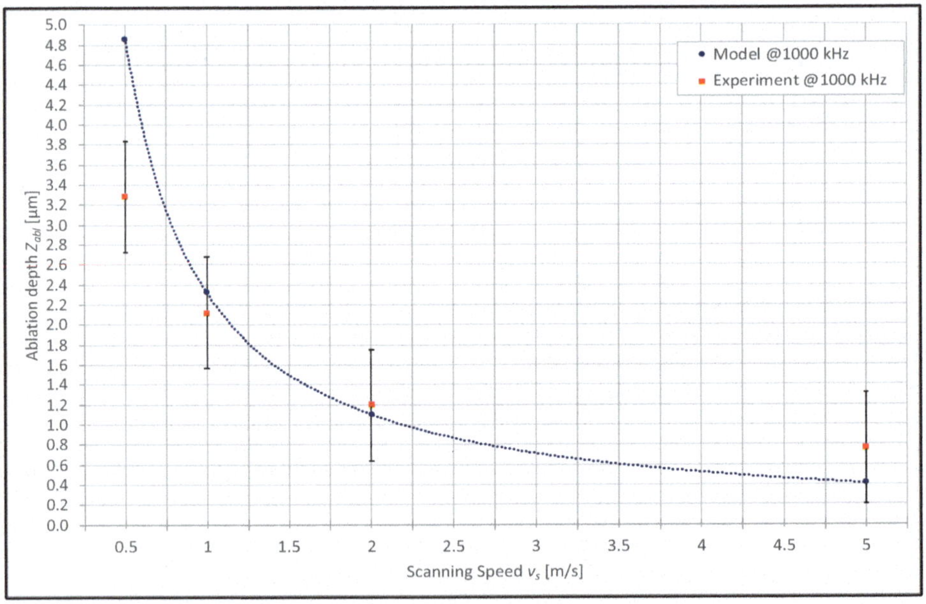

Figure 81: Experimental validation of the multi-pulse picosecond laser beam ablation model with single-pulse on cemented WC at a pulse repetition rate of f = 1000 kHz

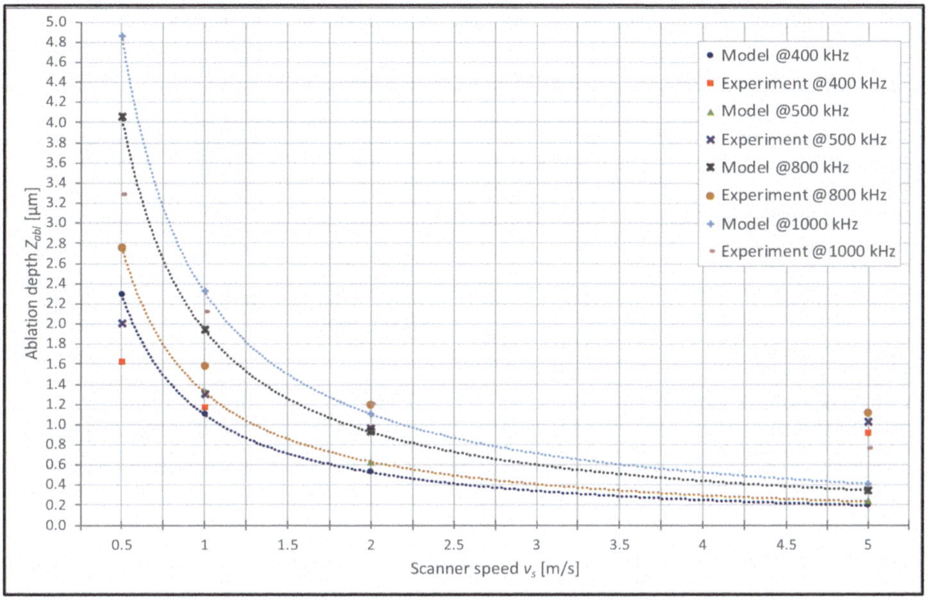

Figure 82: Multi-pulse picosecond laser beam ablation modeling on cemented WC with a burst-mode level of $n = 5$ and their respective experimental validations at all available pulse repetition rates

5.4.4 Multi-pulse laser ablation model with burst-mode

The development of the multi-pulse ablation with burst-mode *BM* basically differs from the model using multi-pulse ablation with single-pulse mode in the fact that every pulse is partitioned in a n number of micro pulses. The product of $N \times n$ affects the behavior of the laser interaction with the material causing a different influence of the incubation effect, and consequently, a different behavior of threshold fluence. The objective of the model with burst-mode is to describe the behavior of picosecond laser beam ablation using burst-mode on cemented WC in terms of ablation depth Z_{abl}. For this analysis, two settings of n are used to gain information; the first, $n = 5$ micro pulses, and the second, $n = 10$ micro pulses, as presented in the summary of the parameters and setting used in the model with burst-mode (table 24).

Table 24: Summary of parameters and settings used in the model with burst-mode

Properties / parameters	Element	Value	Unit	Source
Laser beam power P_L	Laser system	100, 50	[%][W]	Measured
Beam radius ω_0	Laser system	14	[µm]	Measured
Pulse length τ_H	Laser system	10	[ps]	Data sheet
Pulse and burst mode BM	Laser system	5, 10	[-]	Data sheet
Pulse repetition rate f	Laser system	400, 500, 800, 1000	[kHz]	Data sheet
Scanner speed v_s	Scanner system	500, 1000, 2000, 5000	[mm/s]	Data sheet
Incubation factor S	Material	0.75	[-]	Subchapter 5.1
Reflectivity R	Material	0.33	[-]	Subchapter 5.2
Optical penetration depth δ	Material	2.25 x 10^{-8}	[m]	Subchapter 5.3
Threshold fluence $F_{th(1)}$	Material	0.1964	[J/cm^2]	Subchapter 5.3

Threshold fluence $F_{th(1)}$ is taken from the analysis of the Beer-Lambert law and heat transfer presented in subchapter 5.3

Each setting of n affects the total number of pulses irradiated on the material; for this reason, equation 49 determines the number of pulses in burst-mode $N_{(n)}$ irradiating the same area on the surface of the material. Due to the ultra-short separation time of each of the bursts, a burst overlap lower than 99% is neglected, and thus, the pulse overlap PO of the multi-pulse irradiation with single-pulse mode remains the same for multi-pulse irradiation with burst-mode for each set of bursts. The results of the calculation of $N_{(n)}$ are presented in table 25.

The adaptation of N to burst-mode serves as an important element for the calculation of the threshold fluence F_{th} with the use of equation 50 (and the incubation factor S for cemented WC). The following table 26 presents the calculated threshold fluence for $n = 5$ and $n = 10$.

Table 25: Number of pulses on the same area of the surface with single pulse mode, burst-mode $n = 5$ and $n = 10$

Pulse repetition rate	Scanner speed	Number of pulses ♥	Number of pulses with BM ($n = 5$)	Number of pulses with BM ($n = 10$)
f	v_s	N	$N_{(n)}$	$N_{(n)}$
[kHz]	[mm/s]	[-]	[-]	[-]
400	500	22.4	112	224
	1000	11.2	56	112
	2000	5.6	28	56
	5000	2.24	11.2	22.4
500	500	28	140	280
	1000	14	70	140
	2000	7	35	70
	5000	2.8	14	28
800	500	44.8	224	448
	1000	22.4	112	224
	2000	11.2	56	112
	5000	4.48	22.4	44.8
1000	500	56	280	560
	1000	28	140	280
	2000	14	70	140
	5000	5.6	28	56

♥ Calculations for a beam radius ω_0 of 14 μm

Table 26: Modelled threshold fluence F_{th} for $N_{(n)}$

Pulse repetition rate	Scanner speed	Number of pulses with BM ($n = 5$)	Threshold fluence *		Number of pulses with BM ($n = 10$)	Threshold fluence *	
f	v_s	$N_{(n)}$	F_{th}		$N_{(n)}$	F_{th}	
[kHz]	[mm/s]	[-]	[J/m²]	[J/cm²]	[-]	[J/m²]	[J/cm²]
400	500	112	603	0.06	224	507	0.05
	1000	56	718	0.07	112	603	0.06
	2000	28	854	0.08	56	718	0.07
	5000	11.2	1073	0.107	22.4	903	0.09
500	500	140	571	0.05	280	480	0.04
	1000	70	679	0.06	140	571	0.05
	2000	35	807	0.08	70	679	0.06
	5000	14	1015	0.101	28	854	0.08
800	500	224	507	0.05	448	427	0.04
	1000	112	603	0.06	224	507	0.05
	2000	56	718	0.07	112	603	0.06
	5000	22.4	903	0.09	44.8	759	0.07
1000	500	280	480	0.04	560	403	0.04
	1000	140	571	0.05	280	480	0.048
	2000	70	679	0.06	140	571	0.05
	5000	28	854	0.08	56	718	0.07

* Based on the threshold fluence for one pulse $F_{th(1)} = 0.1964$ J/cm²

The average input fluence F irradiated from the laser system is calculated considering the Gaussian distribution, the Beer-Lambert law analysis and the laser and scanner conditions of the system, as done before for single pulse ablation. The use of burst-mode modifies significantly the irradiation of the material by partitioning the pulse into micro pulses. This fact has to be transferred to the calculation of the pulse energy emitted on the material. The fluence of a burst $F_{(n)}$ results from the average input fluence of a pulse F divided by the number of bursts n selected, as described in formula 52.

$$F_{(n)} = \frac{F}{n} \tag{52}$$

Thus, the calculation of $F_{(n)}$ for each of the pulse repetition rates is defined and listed in table 27 for the burst-mode levels of $n = 5$ and $n = 10$, respectively.

Table 27: Average input fluence in the application of the burst-mode

Pulse repetition rate f	Average input fluence F	Average input fluence $F_{(n)}$, $n = 5$	Average input fluence $F_{(n)}$, $n = 10$
[kHz]	[J/m²]	[J/m²]	[J/m²]
400	85423	17084	8542
500	68338	13667	6833
625	54670	10934	5467
800	42711	8542	4271
1000	34169	6833	3416

Beam power P_L of 50 W (100%)
Beam radius ω_0 of 14 µm

The collection of all the required elements targeting ablation depth Z_{abl} with the application of burst-mode is fulfilled and run in equation 51. The modelled results of the behavior of threshold fluence F_{th} and ablation depth Z_{abl} versus scanner speed v_s on cemented WC are illustrated in the following figures 83 to 90 for a burst-mode level of $n = 5$ and $n = 10$, respectively.

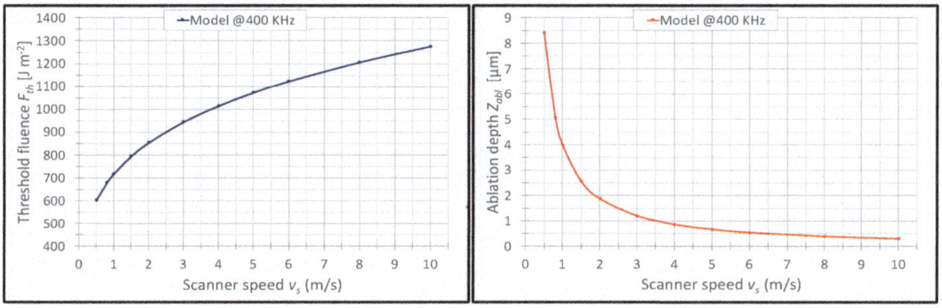

Figure 83: Results of the modeling of threshold fluence F_{th} (left) and ablation depth Z_{abl} (right) with a burst-mode level of $n = 5$ at $f = 400$ kHz vs scanner speed v_s

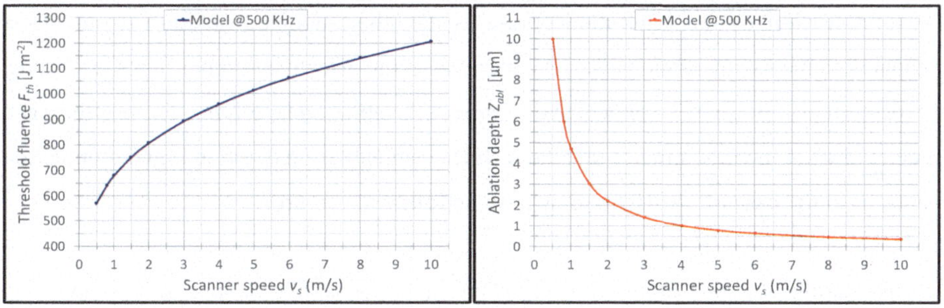

Figure 84: Results of the modeling of threshold fluence F_{th} (left) and ablation depth Z_{abl} (right) with a burst-mode level of $n = 5$ at $f = 500$ kHz vs scanner speed v_s

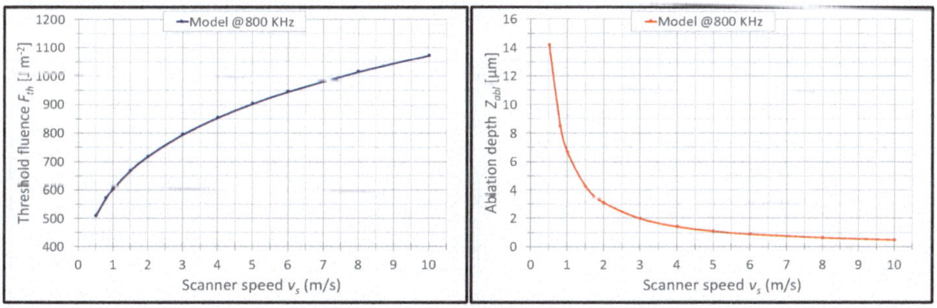

Figure 85: Results of the modeling of threshold fluence F_{th} (left) and ablation depth Z_{abl} (right) with a burst-mode level of $n = 5$ at $f = 800$ kHz vs scanner speed v_s

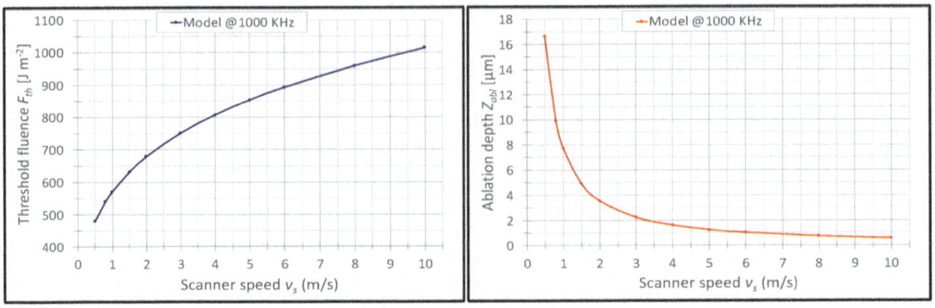

Figure 86: Results of the modeling of threshold fluence F_{th} (left) and ablation depth Z_{abl} (right) with a burst-mode level of $n = 5$ at $f = 1000$ kHz vs scanner speed v_s

The introduction of the burst-mode presents a higher modeled ablation depth when compared to the use of single pulse. The increase of the effective number of pulses hitting the material also presents an expected decrease of the ablation threshold. These effects are more evident in the modeling of threshold fluence and ablation depth in higher number of bursts, as in the case of $n = 10$, as shown in the following figures.

The irradiation of a high number of pulses on the material, like the use of high levels of burst-mode, can produced side-effects, such as plasma shielding, which depend on several experimental conditions. These circumstances were not considered in the model and required a detailed customized analysis.

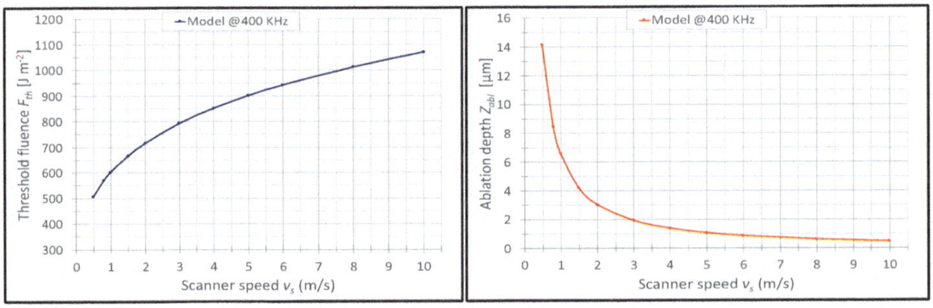

Figure 87: Results of the modeling of threshold fluence F_{th} (left) and ablation depth Z_{abl} (right) with a burst-mode level of $n = 10$ at $f = 400$ kHz vs scanner speed v_s

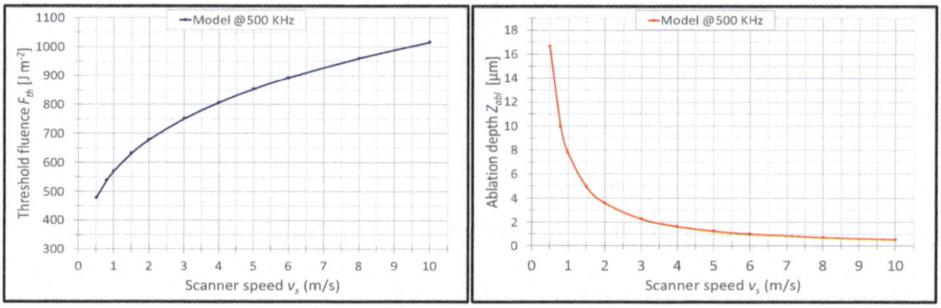

Figure 88: Results of the modeling of threshold fluence F_{th} (left) and ablation depth Z_{abl} (right) with a burst-mode level of $n = 10$ at $f = 500$ kHz vs scanner speed v_s

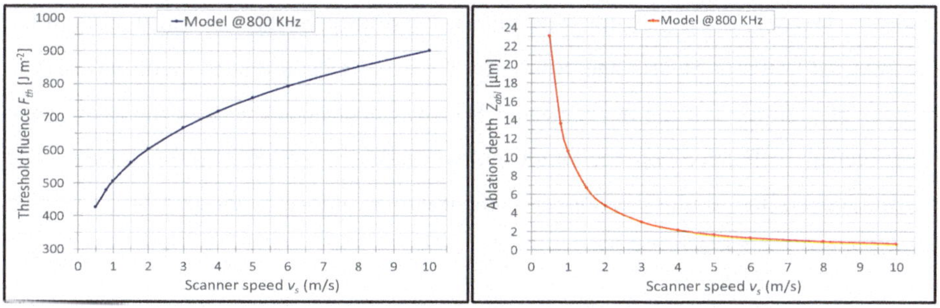

Figure 89: Results of the modeling of threshold fluence F_{th} (left) and ablation depth Z_{abl} (right) with a burst-mode level of $n = 10$ at $f = 800$ kHz vs scanner speed v_s

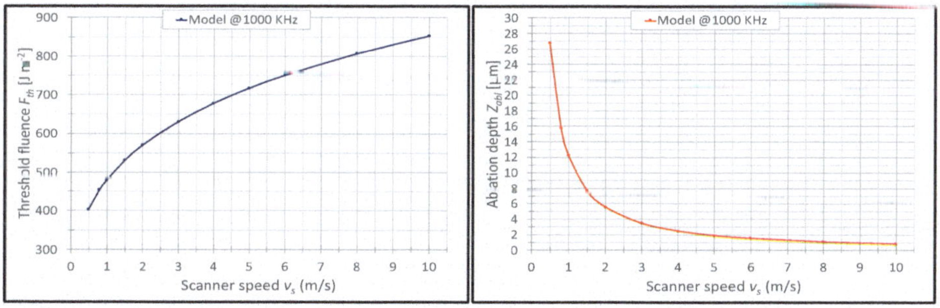

Figure 90: Results of the modeling of threshold fluence F_{th} (left) and ablation depth Z_{abl} (right) with a burst-mode level of $n = 10$ at $f = 1000$ kHz vs scanner speed v_s

The use of a burst-mode level of $n = 10$ in the model offers the highest value of ablation depth when compared to single pulse and to a burst-mode level of $n = 5$. The following subchapter contrasts these modeled results to the experimental findings.

5.4.5. Experimental validation of the multi-pulse model with burst-mode

The validation of the multi-pulse model applying burst-mode levels of $n = 5$ and $n = 10$ was done according to the experimental validation in subchapter 5.4.3. Therefore, the parameters, settings, conditions and set-up for the experiments were kept the same as modeled. The only different setting, as main focus of this subchapter, was the use of the two selected levels of burst-mode. Similarly, the experimental procedure of the ablation of lines of a defined length on cemented WC was kept constant as in the case of the experimental validation of the multi-pulse model with single-pulse mode. The measurement of the results was done with the use of a confocal microscope and targeting Z_{abl}, as done in subchapter 5.4.3.

Firstly, the execution of the validation experiments applying a burst-mode level of $n = 5$ was done and gave the following results after measurement. A comparison of these results with the results of the model can be observed in table 28.

Table 28: Results of the experimentation with a burst-mode level of $n = 5$

Pulse repetition rate	Scanner speed	Number of pulses	Ablation depth	
f	v_s	$N_{(n)}$	Model Z_{abl}	**Experiment Z_{abl}**
[kHz]	[mm/s]	[-]	[µm]	[µm]
400	500	112	8.41	**10.49**
	1000	56	3.99	**6.54**
	2000	28	1.88	**3.23**
	5000	11.2	0.70	**4.27**
500	500	140	9.99	**11.95**
	1000	70	4.72	**6.51**
	2000	35	2.22	**3.16**
	5000	14	0.82	**4.17**
800	500	224	14.20	**22.32**
	1000	112	6.67	**8.75**
	2000	56	3.12	**6.13**
	5000	22.4	1.13	**5.74**
1000	500	280	16.70	**28.43**
	1000	140	7.81	**10.74**
	2000	70	3.63	**6.58**
	5000	28	1.31	**8.50**

The behavior of the ablation depth results Z_{abl} from the model and experiments with a burst-mode level of $n = 5$ is illustrated in the following figures (91, 92, 93 and 94) as a function of the scanner speed v_s and at the four selected pulse repetitions rates f.

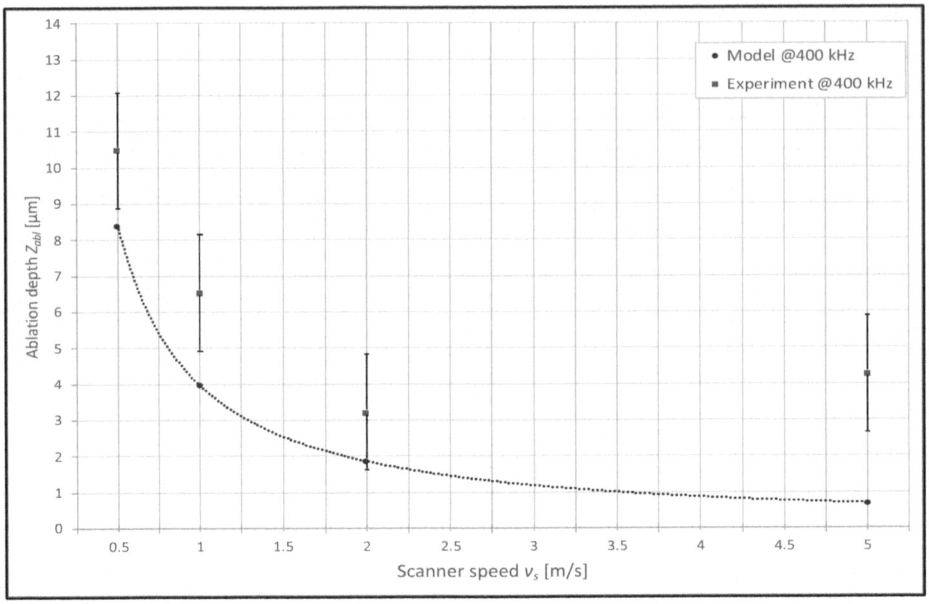

Figure 91: Experimental validation of the multi-pulse picosecond laser beam ablation model with a burst-mode level of $n = 5$ on cemented WC at a pulse repetition rate of $f = 400$ kHz

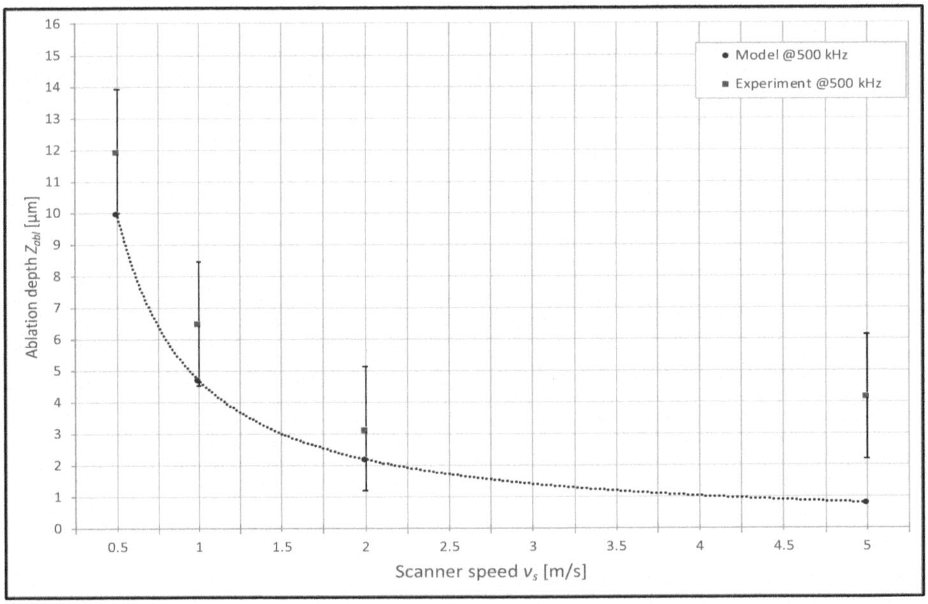

Figure 92: Experimental validation of the multi-pulse picosecond laser beam ablation model with a burst-mode level of $n = 5$ on cemented WC at a pulse repetition rate of $f = 500$ kHz

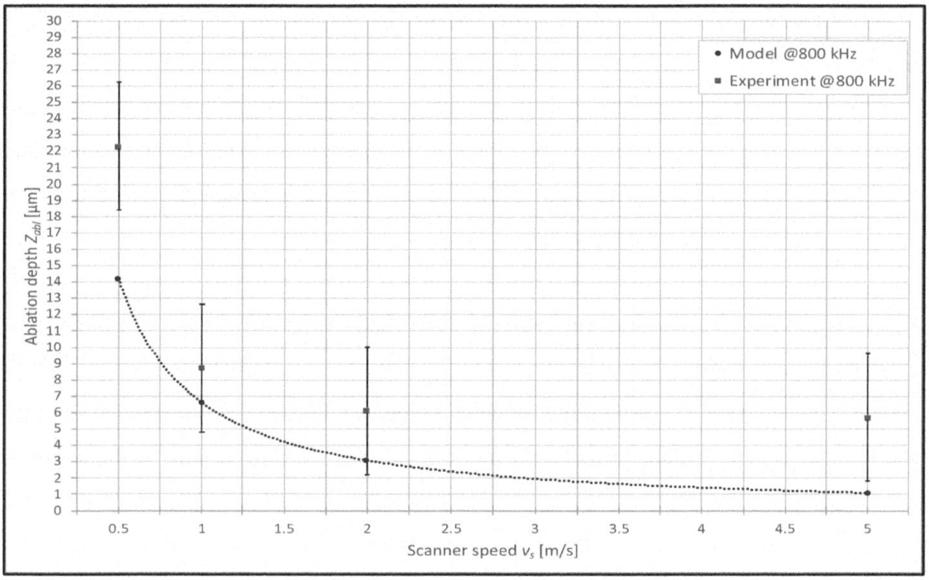

Figure 93: Experimental validation of the multi-pulse picosecond laser beam ablation model with a burst-mode level of $n = 5$ on cemented WC at a pulse repetition rate of $f = 800$ kHz

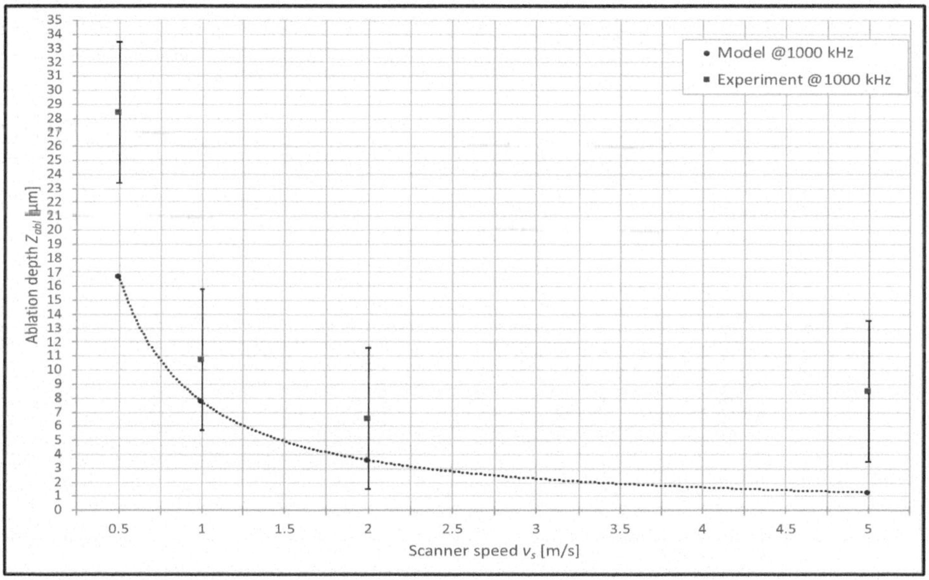

Figure 94: Experimental validation of the multi-pulse picosecond laser beam ablation model with a burst-mode level of $n = 5$ on cemented WC at a pulse repetition rate of $f = 1000$ kHz

A corresponding behavior can be observed in the results of the experiments when compared to the results of the model; the increase of the number of pulses causes in the experimentation higher ablation depth values as modeled but following a harmonic behavior between both of them. An overview of all models and experimental validations at all pulse repetition rates with a burst-mode level of $n = 5$ is presented in figure 95 for comparison. Error is omitted for visualization.

The execution of the experiments using a burst-mode level of $n = 10$ completed the validation experiments of this work. Table 29 lists the resulted ablation depth of the experiments and their corresponding modeled value as a function of pulse repetition rate and scanner speed. Similar to the burst-mode level of $n = 5$, the results of the ablation depth values of the experiments using a burst-mode level of $n = 10$ are plotted and contrasted versus the results of the model as a function of the scanner speed v_s and at the four selected pulse repetitions rates f.

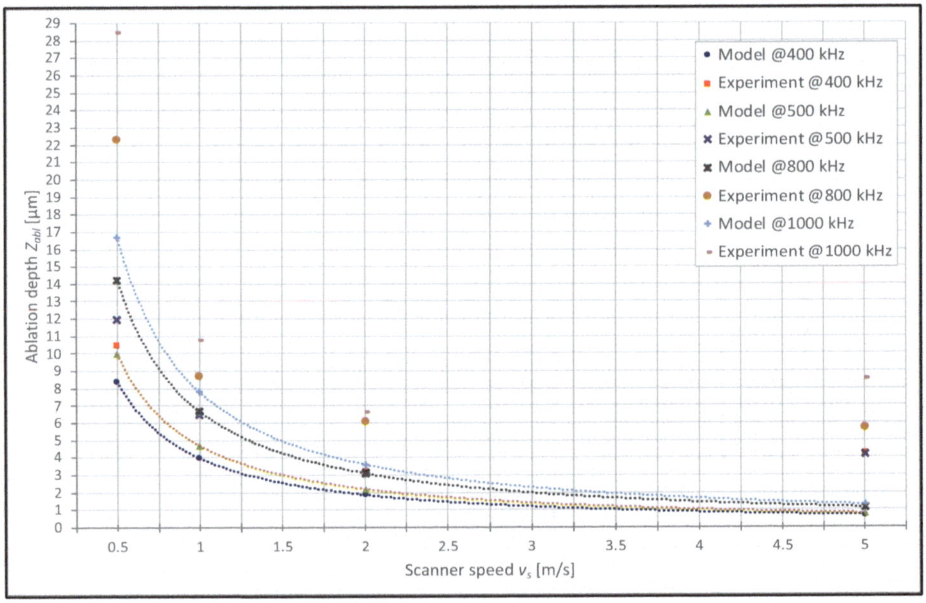

Figure 95: Multi-pulse picosecond laser beam ablation modeling on cemented WC with a burst-mode level of $n = 5$ and their respective experimental validations at all available pulse repetition rates

Table 29: Results of the experimentation with a burst-mode level of $n = 10$

Pulse repetition rate	Scanner speed	Number of pulses	Ablation depth	
f	v_s	$N_{(n)}$	Model Z_{abl}	**Experiment Z_{abl}**
[kHz]	[mm/s]	[-]	[µm]	[µm]
400	500	224	14.20	**26.40**
	1000	112	6.67	**10.60**
	2000	56	3.12	**8.86**
	5000	22.4	1.13	**9.76**
500	500	280	16.70	**29.14**
	1000	140	7.81	**11.61**
	2000	70	3.63	**8.44**
	5000	28	1.31	**7.39**
800	500	448	23.18	**-20.20**
	1000	224	10.72	**22.82**
	2000	112	4.92	**12.36**
	5000	44.8	1.74	**6.20**
1000	500	560	26.87	**-16.95**
	1000	280	12.34	**12.87**
	2000	140	5.63	**10.09**
	5000	56	1.96	**6.87**

Figures 96, 97, 98 and 99 illustrate the resulted behavior of ablation depth as a function of each f = 400 kHz, 500 kHz, 800 kHz and 1000 kHz, respectively. In addition, figures 98 and 99 show a detailed view of the formation of structures and cavities of selected samples.

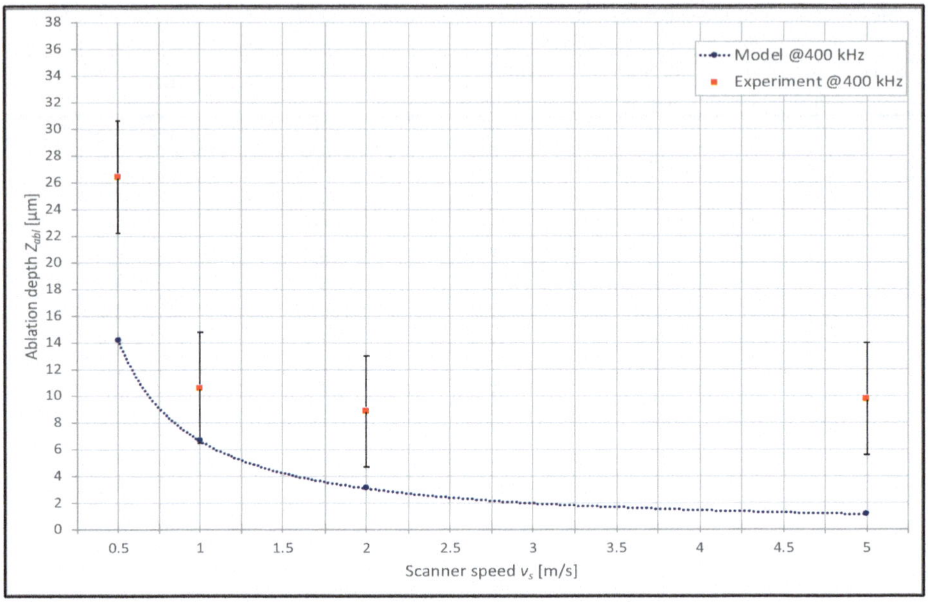

Figure 96: Experimental validation of the multi-pulse picosecond laser beam ablation model with a burst-mode level of $n = 10$ on cemented WC at a pulse repetition rate of $f = 400$ kHz

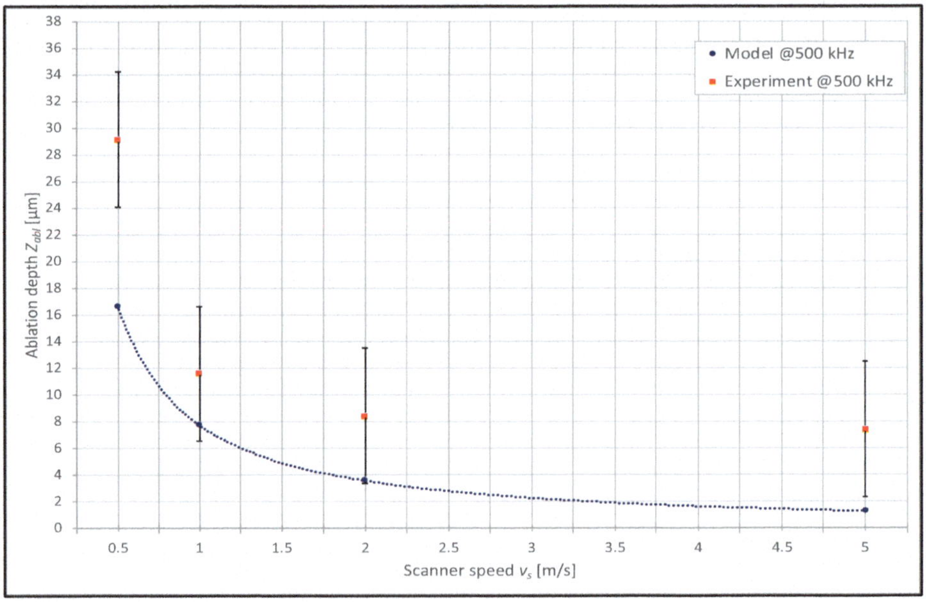

Figure 97: Experimental validation of the multi-pulse picosecond laser beam ablation model with a burst-mode level of $n = 10$ on cemented WC at a pulse repetition rate of $f = 500$ kHz

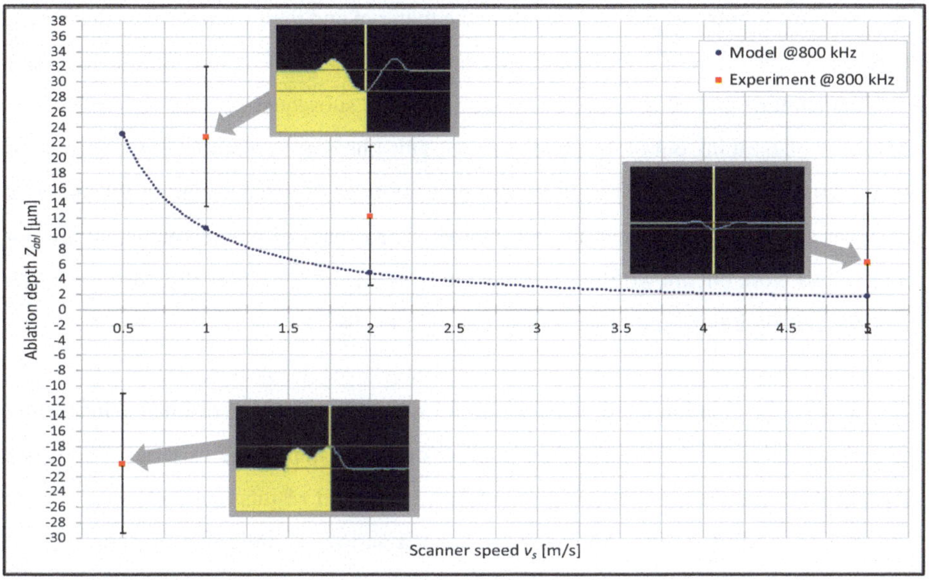

Figure 98: Experimental validation of the multi-pulse picosecond laser beam ablation model with a burst-mode level of n = 10 on cemented WC at a pulse repetition rate of f = 800 kHz

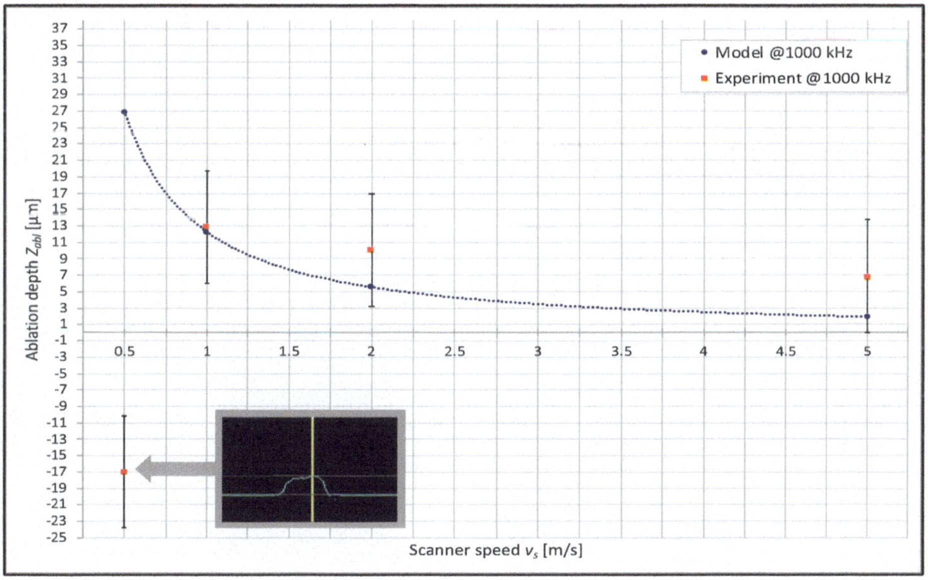

Figure 99: Experimental validation of the multi-pulse picosecond laser beam ablation model with a burst-mode level of n = 10 on cemented WC at a pulse repetition rate of f = 1000 kHz

The rise of the number of pulses irradiating the material and their respective intrinsic high energy, as in the cases of low scanner speeds levels at f = 800 kHz and 1000 kHz causes a high amount of melted material interacting with the subsequent pulses and with a low possibility of escaping out of the ablated zone . This phenomenon produces the formation of a material structure with a height instead of a cavity as depicted in figures 98 and 99 at a scanner speed of v_s = 0.5 m/s in both f cases. The causes of this phenomenon are not focus of this work but are worthy of further study in the fields of laser structuring, plasma behavior and evolution, and process monitoring [144; 145].

The value of the ablation depth behaves uniformly to the modeled results by increasing the scanner speed, i.e., decreasing the number of pulses hitting the material. There is a constant difference among the experimental and modeled results that follows a similar path in both kind of results. An overview for comparison is presented in figure 100; it illustrates all models and experimental validations at all pulse repetition rates with a burst-mode level of n = 10. Error is omitted for visualization.

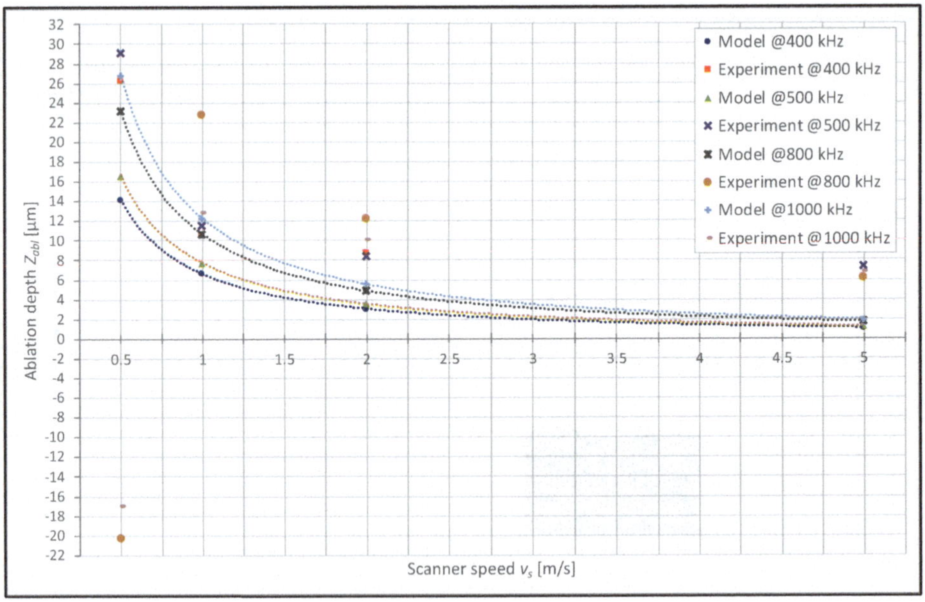

Figure 100: Multi-pulse picosecond laser beam ablation modeling on cemented WC with a burst-mode level of n = 10 and their respective experimental validations at all available pulse repetition rates

The comparison of the results of both burst-mode levels, n =5 and n = 10, at each pulse repetition rate f is presented in the following figures 101, 102, 103 and 104. It can be observed a higher ablation rate Z_{abl} with use of a burst-mode level of n = 10 at both lower pulse repetition rates (f = 400 and 500 kHz) at all scanner speeds. However, high pulse repetition rates (f = 800 and 1000 kHz) cause a reduction of the ablation depth difference among n = 5 and n = 10 and the formation of structures at the lowest scanner speed.

In conclusion, a burst-mode level of n = 5 offers a constant ablation that follows its respective model; On the other hand, n = 10 presents a higher ablation depth value than n = 5 and unexpected phenomena at a scanner speed v_s = 0.5 m/s when combined with high pulse repetition rates (f = 800 and 1000 kHz). The combination of burst-mode levels, number of pulses on the same spot (pulse repetitions rates and scanner speed) and input fluence (laser power and pulse energy) represents a key aspect for an efficient laser beam ablation.

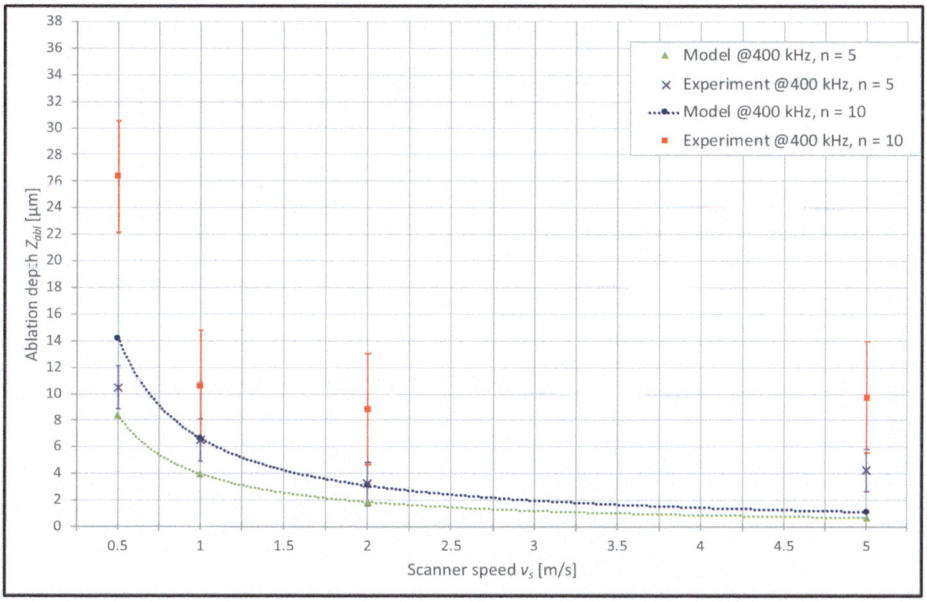

Figure 101: Models and experimental validations of burst-mode levels n = 5 and n = 10 at a pulse repetition rate of f = 400 kHz

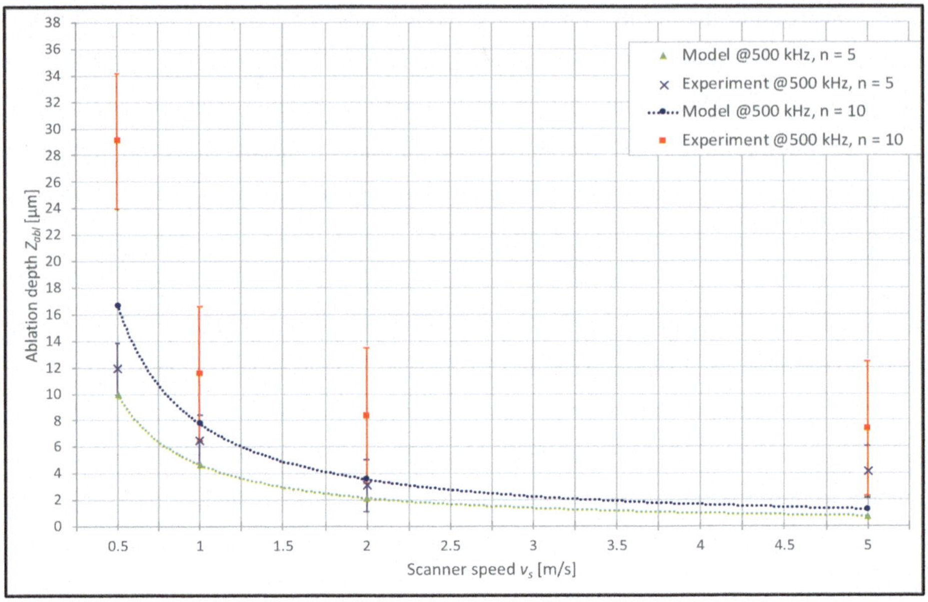

Figure 102: Models and experimental validations of burst-mode levels $n = 5$ and $n = 10$ at a pulse repetition rate of $f = 500$ kHz

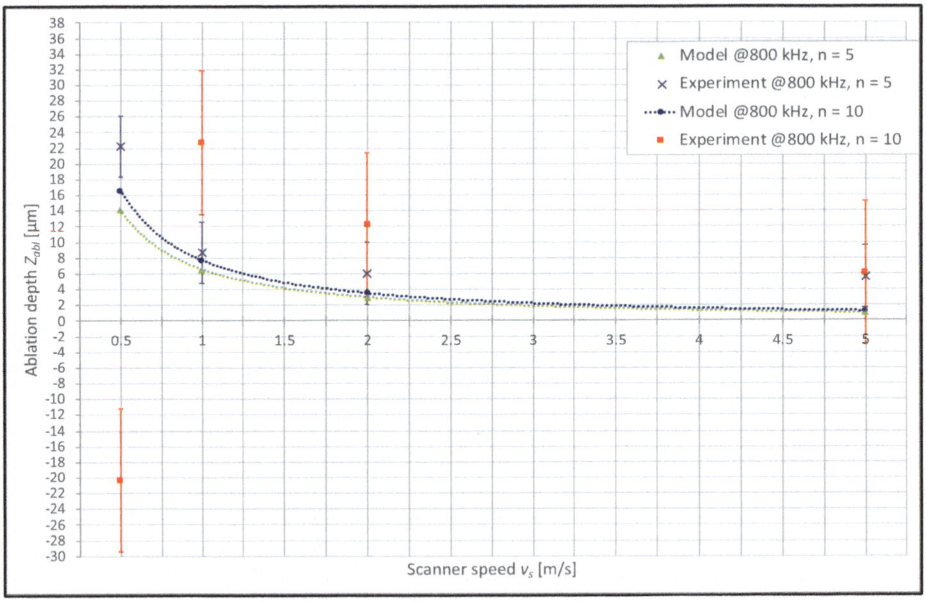

Figure 103: Models and experimental validations of burst-mode levels $n = 5$ and $n = 10$ at a pulse repetition rate of $f = 800$ kHz

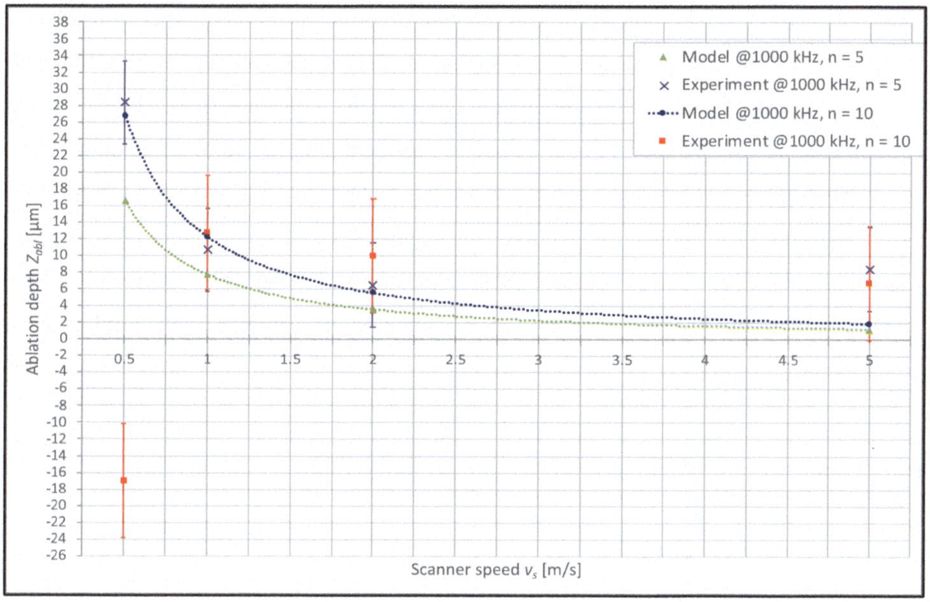

Figure 104: Models and experimental validations of burst-mode levels $n = 5$ and $n = 10$ at a pulse repetition rate of $f = 1000$ kHz

5.4.6 Conclusions

The application of the multi-pulse laser ablation model in this work helped to understand the development of multi-pulse picosecond laser irradiation on cemented WC from a theoretical approach. The fundamental behavior description of the pulse was made in subchapter 5.3 by the Beer-Lambert law analysis, which was the main initial input of the multi-pulse models. The later experimental validation helped to locate the models in their respective significance scope in both modes: multi-pulse laser ablation with single-pulse and multi-pulse laser ablation with burst-mode. The number of pulses on the same area or same spot resulted in a relevant factor to be considered at a constant average laser beam power. Mainly, resulting from the setting of pulse repetitions rate, scanner speed, single and burst-mode, and the choice of optics (focus diameter), the number of pulses represents a controllable parameter of relative ease with a direct relationship to the ablation behavior. Figure 105 introduces a comparison of the experimental ablation depth results to the modeled behavior of ablation depth as a function of the logarithmic number of irradiated pulses on the same area per mode (single-pulse, $n = 5$ bursts and n = 10 bursts). It can be observed a constant relationship among both types of data with a positive difference to the experimental results. Higher number of pulses represent a higher value of ablation depth for the three groups. Only two results, where the highest amount of pulses was present ($N_{(n)} = 448$ and $N_{(n)} = 560$, at $n = 10$ bursts; see negative axis of Z_{abl} in figure 105), showed a generation of a structure or negative ablation (see figures 98 and 99, respectively). This phenomenon represents a boundary for the ablation of cemented WC under those conditions.

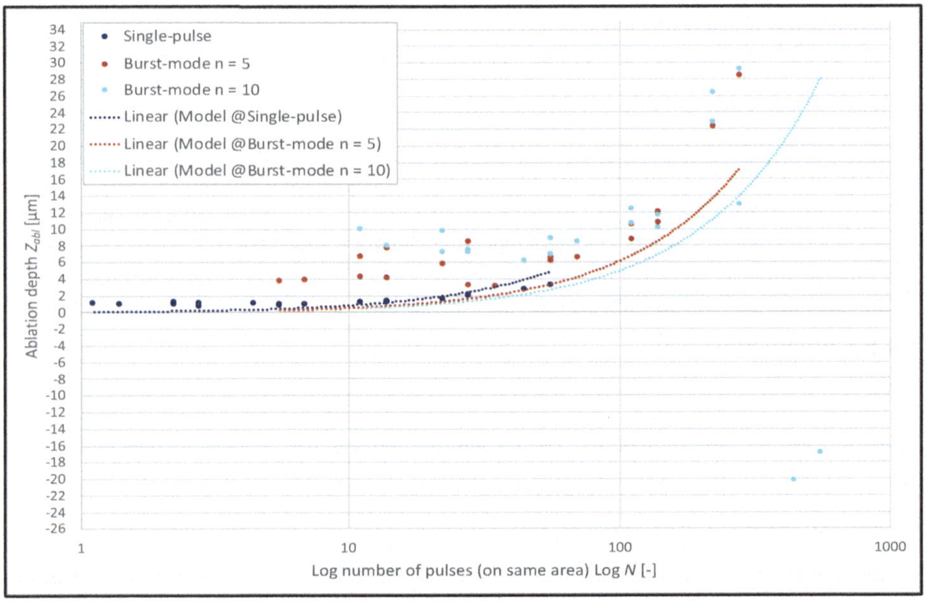

Figure 105: Experimental ablation depth results versus the modeled behavior of ablation depth as a function of the logarithmic number of irradiated pulses on the same area per mode (single-pulse, $n = 5$ bursts and n = 10 bursts)

Low scanner speeds allow the accumulation of a high number of pulses on the same area; the selection of the highest pulse repetition rate causes the generation of the mentioned structures. The following figures present the behavior of ablation depth as a function of pulse repetition rate to observe the increment of pulses at a constant low scanner speed (v_s = 0.5 m/s). Figure 106 illustrates the behavior of ablation depth using single-pulse mode.

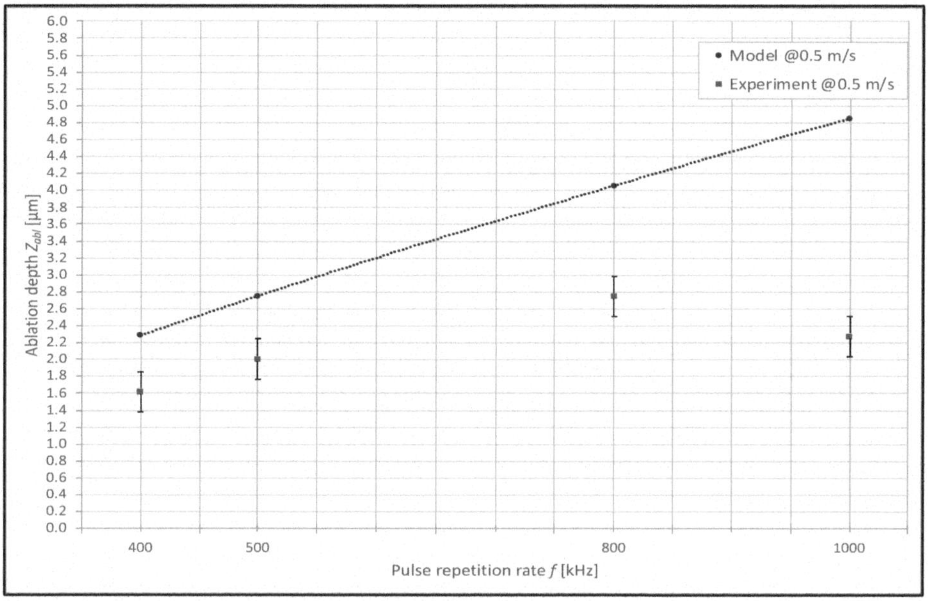

Figure 106: Experimental validation of the multi-pulse picosecond laser beam ablation model with single-pulse on cemented WC as a function of pulse repetition rate f and at a scanner speed of $v_s = 0.5$ m/s

The ablation is constantly lower as expected in the model and with a more noticeable decrease at $f = 1000$ kHz due to the increment of pulses and presumably to the energy accumulation and plasma interactions. On the other hand, figure 107 using a burst mode at $n = 5$ presents a higher ablation depth as expected in the model and with a positive behavior from $f = 800$ kHz to 1000 kHz. This positive ablation can be a hint to better energy distribution on the material thanks to the use of burst-mode, i.e., increasing the number of pulses by N x n and dividing the average input fluence as in equation 52. As a matter of fact, the behavior of experimental results follows the expected model behavior with the exception of the cases with the highest number of pulses.

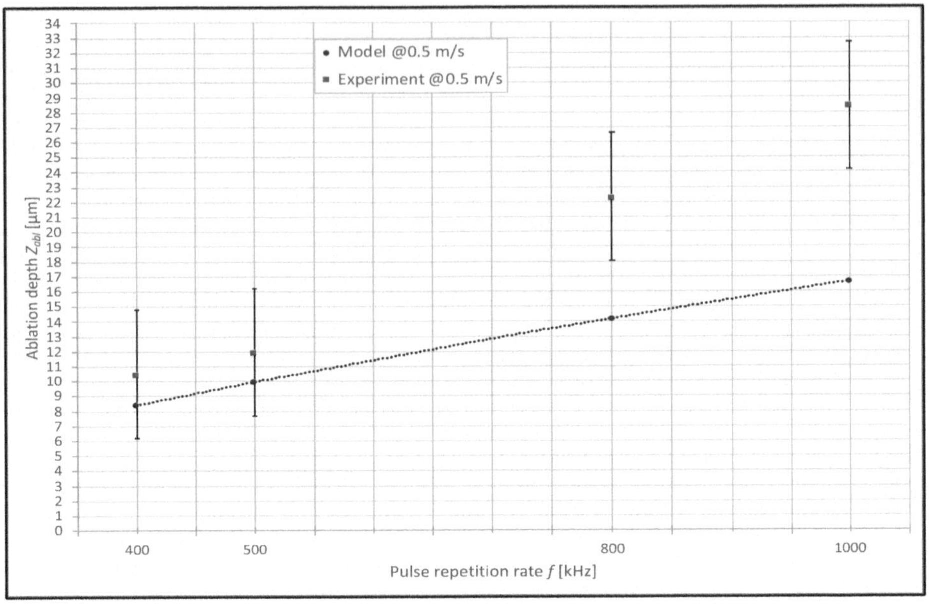

Figure 107: Experimental validation of the multi-pulse picosecond laser beam ablation model with a burst-mode level of $n = 5$ on cemented WC as a function of pulse repetition rate f and at a scanner speed of $v_s = 0.5$ m/s

Contrasting to the behavior of the model but understood for the analysis mentioned before, figure 108 shows the boundary where the two cases with the highest number of pulses at pulse repetition rates $f = 800$ kHz and $f = 1000$ kHz, respectively, cause a drastic generation of a structure or negative ablation. Before that, the behavior under 400 kHz and 500 kHz seems to follow the model. In contrast to the results of $n = 5$ (figure107), the results of $n = 10$ (figure108) only add a higher partition of the pulses and their corresponding input fluence. This causes an input fluence excess on the material that blocks ablation at the highest pulse repetition rates.

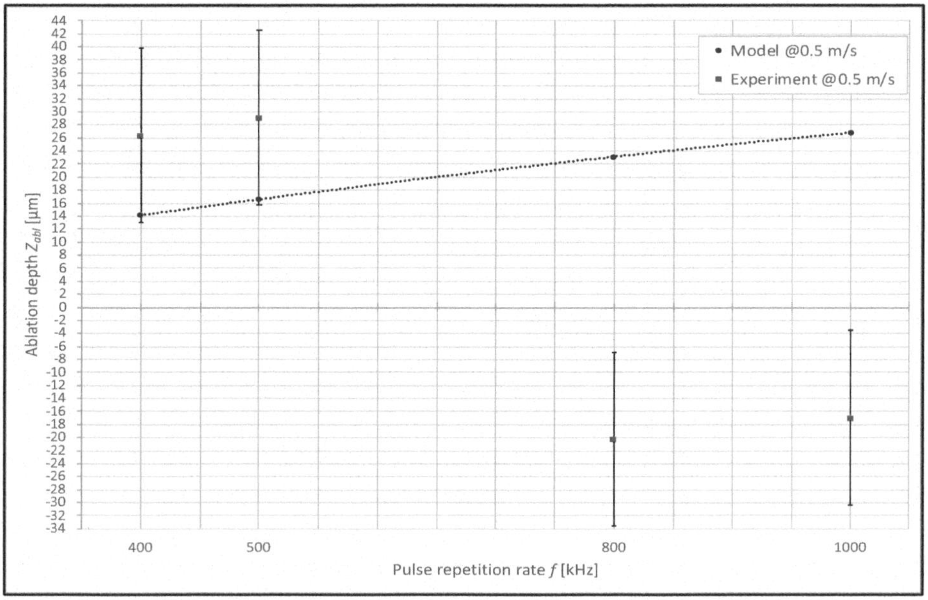

Figure 108: Experimental validation of the multi-pulse picosecond laser beam ablation model with a burst-mode level of $n = 10$ on cemented WC as a function of pulse repetition rate f and at a scanner speed of $v_s = 0.5$ m/s

In the cases of higher mid scanner speeds, e.g., 1 m/s, 2 m/s and 5 m/s, the behavior follows the model, but it shows peculiarities related to the mode of the pulse. The use of single-pulse mode, as in figure 109, presents an accumulation of energy that might cause no gain in the ablation and a slight decrease at higher frequencies, i.e., high number of pulses on the area. Similarly, this behavior is present with the use of burst-mode with a high level of burst, like in the case of $n = 10$. The application of $n = 5$ bursts causes an effective ablation with a slight rise of the ablation depth values at higher pulse repetitions rates, as depicted in figure 110. The application of higher scanner speeds, e.g., from 5 m/s to 10 m/s, shows similar behavior among them. They present a similar behavior to the model but with the presence of a constant difference. Moreover, they present similar decrease and rise of the ablation depth levels at higher pulse repetition rates with the use of single-pulse and burst-mode, respectively.

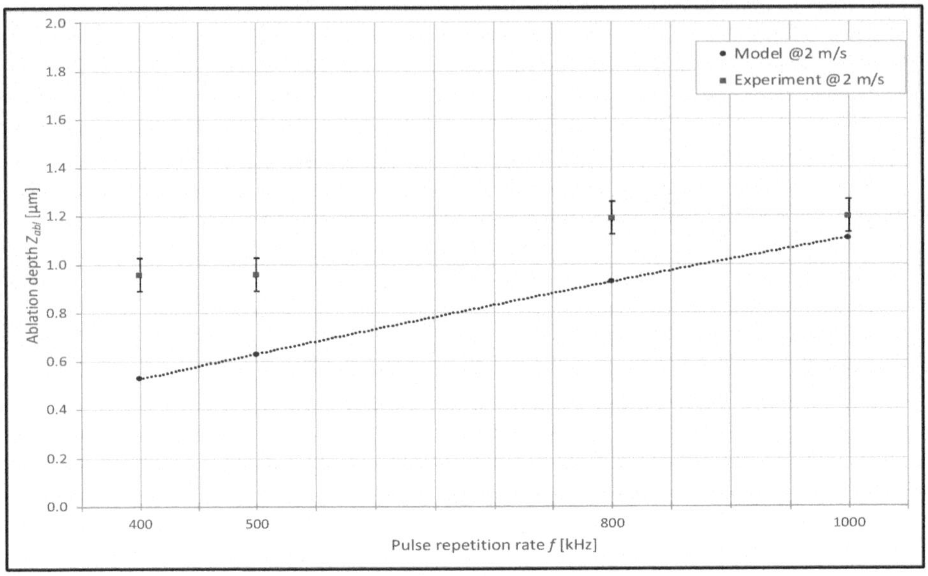

Figure 109: Experimental validation of the multi-pulse picosecond laser beam ablation model with single-pulse on cemented WC as a function of pulse repetition rate f and at a scanner speed of $v_s = 2$ m/s

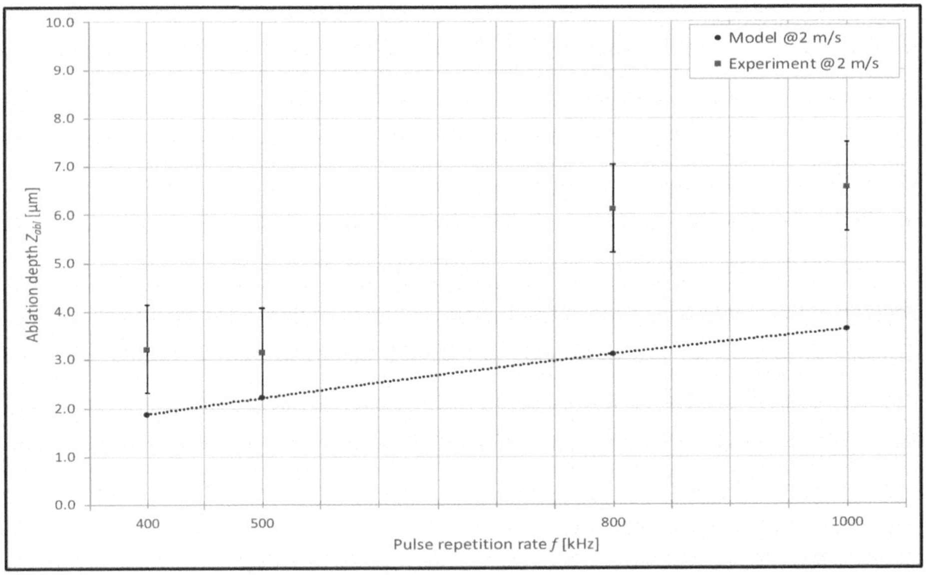

Figure 110: Experimental validation of the multi-pulse picosecond laser beam ablation model with a burst-mode level of $n = 5$ on cemented WC as a function of pulse repetition rate f and at a scanner speed of $v_s = 2$ m/s

Figure 111 is an example of the use at high scanner speeds. It can be observed a similar model behavior at f = 400 kHz and f = 500 kHz and a rise at f = 800 kHz and f = 1000 kHz. As expected, the lower number of pulses at higher scanner speeds reduces the value of ablation depth for all cases.

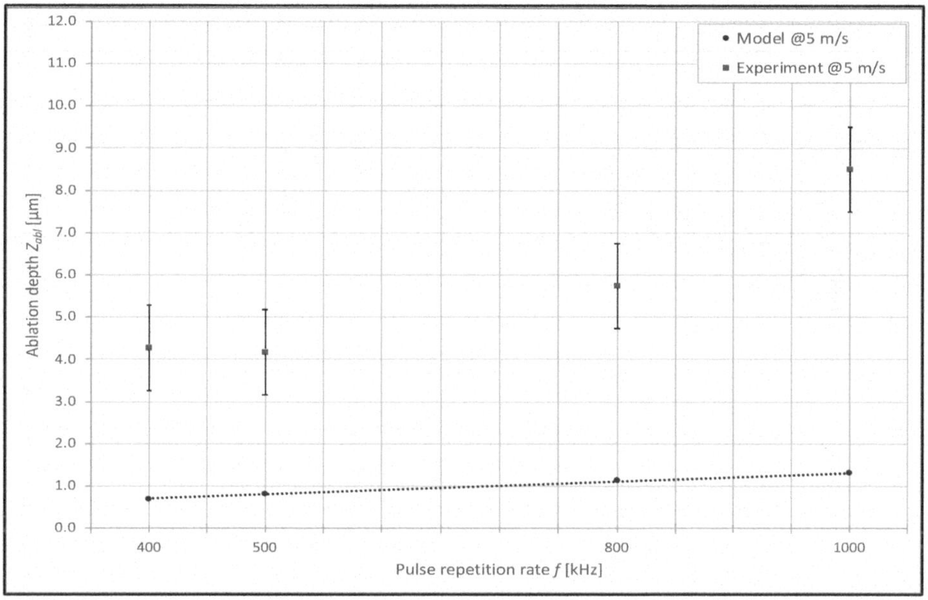

Figure 111: Experimental validation of the multi-pulse picosecond laser beam ablation model with a burst-mode level of n = 5 on cemented WC as a function of pulse repetition rate f and at a scanner speed of v_s = 5 m/s

An important approach to evaluate the relationship between the model and the experiment results is the evaluation of their difference or error. Taking as a point of reference the experimental results, the evaluation of the error of the model can be described in percentage as in equation 53

$$Er_m = \frac{Z_{abl(model)} - Z_{abl(experiment)}}{Z_{abl(experiment)}} \cdot 100 \qquad (53)$$

Where, Er_m is the error of the model when compared to the experiment, $Z_{abl(model)}$ represents the ablation depth results of the model and $Z_{abl(experiment)}$ corresponds to the ablation depth results of the experiment. The following figure illustrates the error of the model Er_m as a function of the number of pulses irradiated on the same area for all modes (figure 112). It can be mainly observed the proportion of the model results vs the experiment under the x axis suggesting higher values of ablation depth from the experiments.

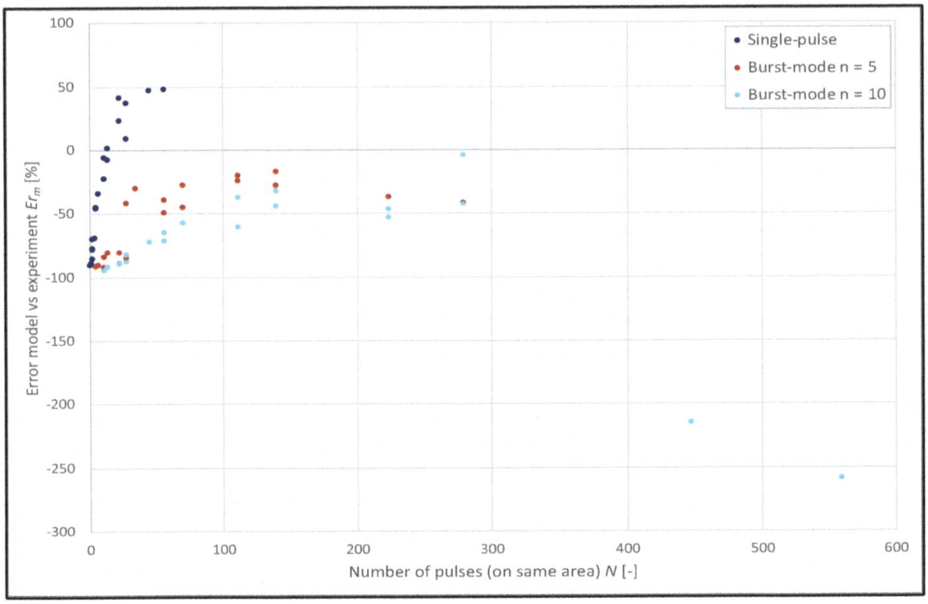

Figure 112: Error of the model Er_m as a function of the number of pulses N irradiated on the same area for all modes

Likewise, it can be observed greater negative values of error with the use of a smaller number of pulses on the same area, stating that there is a higher difference among model and experiment with lower N and $N_{(n)}$. The use of single pulse is balanced in both sides of the error axis. The level of $n = 5$ bursts presents the more constant behavior of the three mode options and the level of $n = 10$ bursts shows mainly higher values of ablation depth in the experiments. As mentioned before, the level of $n = 10$ bursts shows the formation of structures (negative ablation) with a considerable difference of the value of ablation depth to the model.

Figure 113 illustrates the behavior of the model error as a function of the ablation rate value measured from the experiments. The same behavior showing a greater ablation depth value when compared to the model can be observed. Single-pulse mode shows a constant error above and under the x axis and the results using burst-mode maintain a constant higher experimental value. Similarly, it can be differentially observed the depth reached by each of the mode groups, being the groups applying partition of pulses the results with the deepest ablation in the material. A significant difference among $n = 5$ and $n = 10$ in ablation depth cannot be found; a remark, ruling out the structure formation cases, can be the fact that the results of $n = 10$ have their lowest value of Z_{abl} above all single-pulse mode results and some of $n = 5$ (6.2 μm depth).

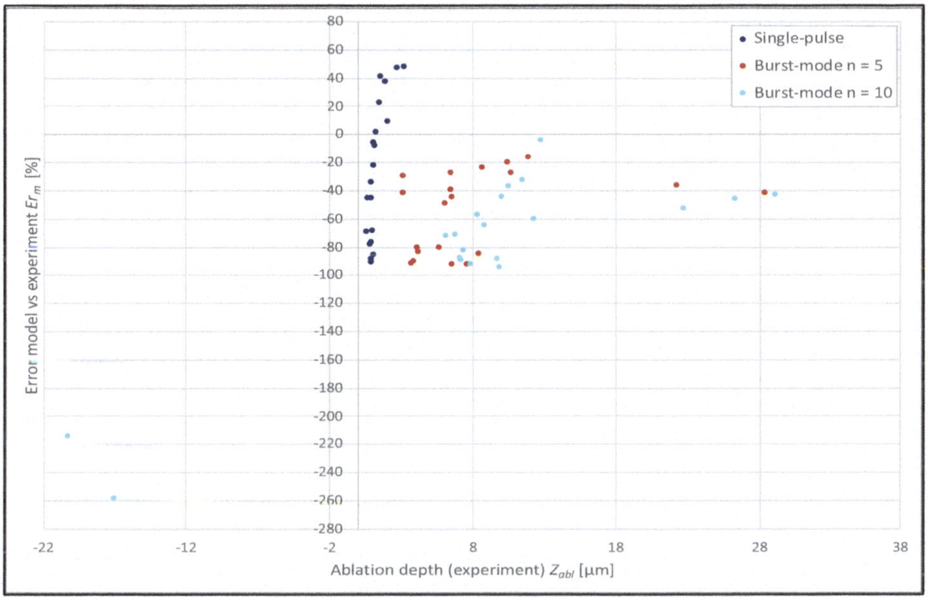

Figure 113: Error of the model Er_m as a function of ablation depth Z_{abl} from the experimental results for all modes

The last error analysis corresponds to the evaluation of the error as a function of the ratio of the input fluence and the threshold fluence F/F_{th}. The ratio targets the understanding of the input and required energy to ablate the material and, basically, how much energy is redundant. Thus, values of the ratio close to 0 represent values with more precise fluence corresponding to the threshold, and values further to 0 represent the values with an excess of input fluence. Negative values of the ratio F/F_{th} are not possible due to the fact that ablation would not take place.

Figure 114 illustrates the error of the model Er_m as a function of the ratio of fluence F and threshold fluence F_{th} for all modes. It can be observed that the group irradiated with single-pulse mode presents the values with the highest ratio, i.e., there is an excess of input fluence form the system, and relatively balance error. On the other hand, the groups working with burst-mode show results with an efficient supply of fluence, being $n = 10$ the most efficient. This efficiency is achieved thanks to the partition of the energy of a single pulse into the number of bursts by the application of the burst-mode.

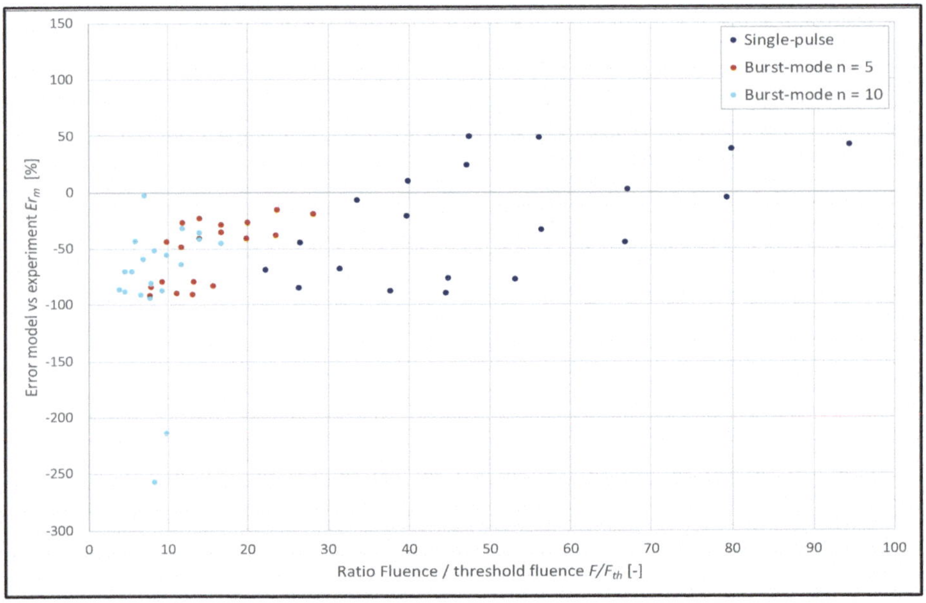

Figure 114: Error of the model Er_m as a function of the ratio of fluence F and threshold fluence F_{th} for all modes

A similar behavior of the experimental results follows the model with the understanding of the variations on certain cases and the difference in ablation depth among both type of results. Constant behaviors under same conditions of scanner speed and pulse repetition rate have been observed in each of the three modes. Boundaries of positive and negative ablation can be observed in the use of a burst-mode level of $n = 10$ that might help in the process of laser structuring. In general, single-pulse mode ablation depth results presented a behavior corresponding to the model and only at the highest pulse repetition rate (1000 kHz) showed a slight decrease of the ablation. The use of burst-mode with $n = 5$ also presented a corresponding model behavior with the exception at high repetitions rates (800 kHz and 1000 kHz), where it presented an increment in the values of ablation. Lastly, burst-mode with a $n = 10$ gave, the already mentioned, structures at low scanner speeds and from mid to high pulse repetition rates presents a rise of the ablation depth values. A future objective has to be to increase the accuracy of the model by integrating more parameter; however, a calibration of it can be done after the understanding of the behavior presented in this work. A valid and practical use can be adapted to other application environments of laser beam ablation.

5.5 Summary of the analytical approach

The path of the analytical approach began with the identification of important information for the analysis. Material and system specification, experience from the experimental approach from this work, search of not in the literature available data served as basis for the approach. Thus, the determination of the incubation factor was a relevant part of the chapter to define, by experimental means based of a known method, the value of the S factor. The result gave a value of an incubation factor in the range of $S = 0.7\text{-}0.8$ ($S = 0.75$ for practical use). This value was used posteriorly in the application of the model. Similarly, there was the need to integrate the value of reflectivity. An experimental set-up was planned and reflection experiments under similar conditions of the laser system on cemented WC and a reference material were executed. A reflectivity value of $R = 0.33$ was found and included in the further analysis of the study. Once all information was investigated and collected, the ablation model using a picosecond laser on cemented WC was developed considering the actual parameters and settings of the picosecond laser beam system. Based on a Beer-Lambert law analysis, the behavior of a single pulse penetrating the material was described. Among other information, the identification of power density and threshold fluence for a single pulse was obtained. Figure 115 illustrates the development of power density *Gt* of a picosecond pulse through cemented WC as function of time *t* and depth *y* into the material.

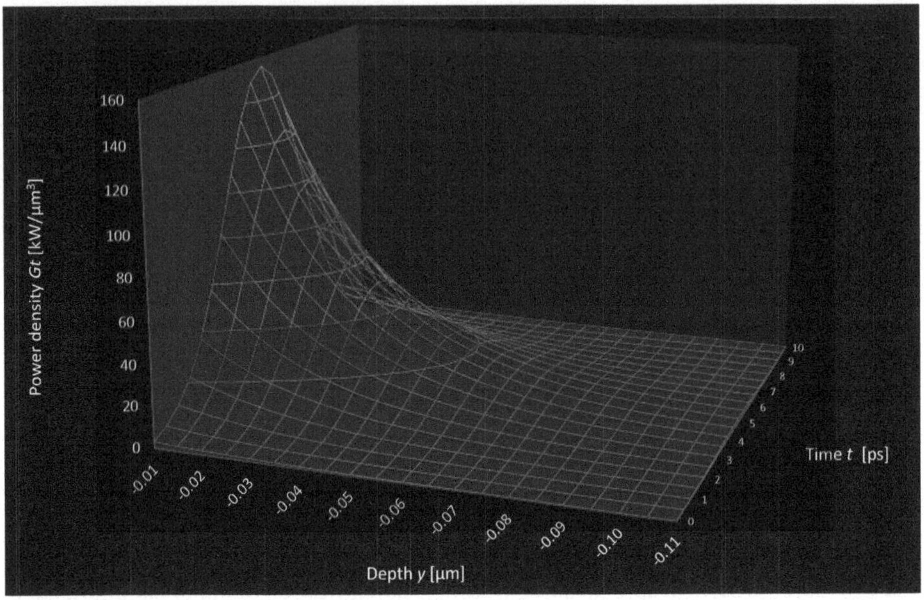

Figure 115: Power density distribution *Gt* of a picosecond laser pulse at 1000 kHz and $P_L = 100\%$ as a function of pulse length τ_H and penetration depth *y*

It can be observed a high-power density on the surface of the material that diminishes through the depth; similarly, it can be seen the Gaussian behavior of the pulse through the pulse length.

Thus, this understanding of the energy of a single pulse going through cemented tungsten carbide served as the basis for the multi-pulse picosecond laser beam ablation model. Therefore, in the multi-pulse investigation, ablation depth Z_{abl} was targeted and the model was executed giving a series of results in four pulse repetition rates, four scanner speeds and three pulse modes. The model results were followed by an experimental validation executed in 5.4.3 and 5.4.5 and discussed in 5.4.6. The comparison of both types of results presented a corresponding behavior with mainly higher ablation depth values in the experimental results (single-pulse mode at low scanner speeds showed a lower ablation depth value than higher scanner speeds and burst-mode) and with the presence of relevant unexpected phenomena, like the cases of structures formation. The use of burst-mode presented higher values of ablation depth and showed an efficient pulse energy distribution in the ratio F/F_{th} even when a high number of pulses could cause an accumulation of energy and thus minimize ablation.

To conclude, the following table 30 summarizes the main parameters, settings and results that were obtained in the multi-pulse laser ablation analysis for all modes. Among the results, it can be found the values registered from the model and experiments with their corresponding number of pulses and threshold fluence for comparison and understanding.

Table 30: Summary of results and corresponding parameter and settings from the multi-pulse laser ablation analysis

Parameters			Single-pulse mode				Burst-mode $n = 5$				Burst-mode $n = 10$			
f	v_s	PO	N	F_{th}	Z_{abl} (model)	Z_{abl} (experiment)	$N_{(n)}$	F_{th}	Z_{abl} (model)	Z_{abl} (experiment)	$N_{(n)}$	F_{th}	Z_{abl} (model)	Z_{abl} (experiment)
[kHz]	[mm/s]	[-]	[-]	[J/cm²]	[µm]	[µm]	[-]	[J/cm²]	[µm]	[µm]	[-]	[J/cm²]	[µm]	[µm]
400	500	0.96	22.4	0.09	2.29	1.62	112	0.06	8.41	10.49	224	0.05	14.20	26.40
	1000	0.91	11.2	0.107	1.10	1.17	56	0.07	3.99	6.54	112	0.06	6.67	10.60
	2000	0.82	5.6	0.12	0.53	0.96	28	0.08	1.88	3.23	56	0.07	3.12	8.86
	5000	0.55	2.24	0.16	0.20	0.92	11.2	0.107	0.70	4.27	22.4	0.09	1.13	9.76
500	500	0.96	28	0.08	2.76	2.01	140	0.05	9.99	11.95	280	0.04	16.70	29.14
	1000	0.93	14	0.101	1.32	1.30	70	0.06	4.72	6.51	140	0.05	7.81	11.61
	2000	0.86	7	0.12	0.63	0.96	35	0.08	2.22	3.16	70	0.06	3.63	8.44
	5000	0.64	2.8	0.15	0.24	1.03	14	0.101	0.82	4.17	28	0.08	1.31	7.39
800	500	0.98	44.8	0.07	4.06	2.75	224	0.05	14.20	22.32	448	0.04	23.18	-20.20
	1000	0.96	22.4	0.09	1.94	1.58	112	0.06	6.67	8.75	224	0.05	10.72	22.82
	2000	0.91	11.2	0.107	0.93	1.19	56	0.07	3.12	6.13	112	0.06	4.92	12.36
	5000	0.78	4.48	0.13	0.35	1.11	22.4	0.09	1.13	5.74	44.8	0.07	1.74	6.20
1000	500	0.98	56	0.07	4.86	3.28	280	0.04	16.70	28.43	560	0.04	26.87	-16.95
	1000	0.96	28	0.08	2.32	2.12	140	0.05	7.81	10.74	280	0.048	12.34	12.87
	2000	0.93	14	0.101	1.11	1.20	70	0.06	3.63	6.58	140	0.05	5.63	10.09
	5000	0.82	5.6	0.12	0.41	0.76	28	0.08	1.31	8.50	56	0.07	1.96	6.87

♥ Calculations for a beam radius ω_0 of 14 µm

♦ Calculations for a pulse length τ_H of 10 ps

* Based on the threshold fluence for one pulse $F_{th(1)} = 0.1964$ J/cm²

Where, f is the pulse repetition rate, v_s corresponds to the scanner speed, PO is the pulse overlap, N represents the number of pulses on the same area, $N_{(n)}$ corresponds to the number on pulses on the same area in burst-mode, F_{th} is threshold fluence, $Zabl_{(model)}$ corresponds to the ablation depth results from the model and $Zabl_{(experiment)}$ represents the ablation depth results from the experiments

6 Evaluation of results and analysis of influential parameters

After the experimental and theoretical investigation in chapter 4 and 5, a model considering all described parameters was defined. A further experimental validation was done releasing results and conclusions of the model. This information output is evaluated and analyzed in this chapter with the objective of consolidating the experience in the form of a framework that contains the methodology and all required information for a further practical application of the work. Thus, chapter 6 is situated in the phases *check* and *do*, which are understood, as the evaluation and application fields of the relevance of the work (figure 4 situates this idea in the structure of the thesis).

The following lines present two parts; the first part corresponds to the evaluation and analysis of the conclusions of the experimental and analytical approaches. The second part introduces the general framework of the work, which is intended to serve as a guideline for the practical use of the work. It contains a methodology flow that emphasizes the procedures and important aspects to apply the model and get relevant information of future laser beam ablation processes with other materials.

6.1 Evaluation and analysis

A fundamental part of this work was the methodology for the identification of the influential parameter in the picosecond laser beam ablation of cemented WC. The objective was the identification of the parameters and corresponding settings affecting significantly the ablation rate. The results presented the cluster of pulse overlap and track overlap as the most significant parameter in the maximization of the ablation rate. Secondly, laser beam power affected ablation significantly, and lastly, the application of burst-mode was found as the third influential factor with an opposite behavior as expected. After the analytical approach, it is clearer to understand the importance of these three parameters. Pulse overlap and track overlap play an important role in the interaction of pulse repetition rate, scanner speed and number of pulses irradiated on the same area; this

J. P. Calderón Urbina, *Efficient material laser beam ablation with a picosecond laser*, Light Engineering für die Praxis, https://doi.org/10.1007/978-3-662-61886-8_6

last parameter plays an important role in how the material is irradiated and exposed to the laser. Laser beam power represents a controllable parameter that can be easily adjusted in the system and can work as an element for acceleration of the process or fine-tuning processing when the system reaches its limits in terms of pulse repetition rate and scanner speed. The use of the last most influential, burst-mode, reveals two important aspects. The first is the increase of the number of pulses on the same area by its ultra-short separation time in nanosecond range. This allows to increase the number of pulses when it is required by partitioning a pulse. The second aspect corresponds to the division of the input fluence by the number of bursts. This fact makes possible a distribution of the pulse energy closest to the threshold fluence of the material, and thus, to avoid an excess of irradiation that might cause thermal affected zones, plasma interactions and irregularities. Burst-mode also represents a parameter easy to integrate from the laser system; it can be set up to 10 bursts, and thus, to facilitate the definition of the desired input fluence.

The second fundamental part of this work was the analytical approach. A key aspect of this approach is the collection of material and system information for the development of a model to describe ablation. Therefore, accurate data is required from different perspectives of material and laser system. Thermal, physical and optical properties of the material are needed in the different phases of the methodology. Literature information is available, but it does not complete all the required information for the investigation. For this reason, the execution of experimental methodologies, such as the determination of the incubation factor and reflectivity in 1064 nm for this work, are important for the accomplishment of the calculations. It is important to execute these experimental methodologies under controlled experimental set-ups and using the same or similar system conditions as the ones targeted in the system under study. The incubation factor of S = 0.75 for cemented WC is located in a similar range among other reference factors; for example, the incubation factor for tungsten (W) is of $S = 0.84$ and the factor for stainless steel of $S = 0.86$ [101; 137]; this fact places the finding as an acceptable value for cemented tungsten carbide. The calculation of the optical penetration depth at λ = 1064 nm in cemented WC gave a value of $\delta = 2.25 \times 10^{-8}$ m necessary for further calculations. Furthermore, the reflectivity factor of the material with an infrared beam was determined to be $R = 0.33$. A reference value of $R = 0.67$ for steel was similarly determined under the same experimental set-up and conditions; this value correspond to the acceptable range found in the literature of other types of steel [139].

Once fulfilling the information requirements and thus based on the material properties of the cemented WC sample, a heat transfer analysis was executed. This analysis conducted a theoretical transformation of the material through its three phases from the solid-state to the vapor-state by calculating the required energy to achieve such phase changes. The resulting required power to evaporate solid cemented WC was presented as the ablation threshold and calculated as threshold fluence. The value of the threshold fluence for one pulse was determined as $F_{th(1)}$ = 0.1964 J/cm^2. The resulting energy value in terms of power density assisted the theory of the Beer-Lambert law for the determination of the behavior of the input power through the material when one picosecond pulse is irradiated under the conditions of the actual laser system. Subsequent to the analysis of one pulse, a multi-pulse laser ablation model was targeted. The description of the ablation behavior under single-pulse mode and burst-mode for four pulse repetition rates and four scanner speeds, as already mentioned in this work, was theoretically done. Posteriorly, each of the three investigation groups (single-pulse, $BM\ n$ = 5 and $BM\ n$ =10) was experimentally validated and the results carefully observed. Conclusions and error analyses were performed in subchapter 5.4.6 and 5.5 concluding as relevant findings to optimize. Once again, it can be said that the number of pulses on the same area in single-pulse mode and with burst-mode plays an important role in how the energy is delivered on the material and consequently in the behavior of ablation depth in multi-pulse ablation. Therefore, the interaction of pulse repetition rate, scanner speed, optics (focal diameter) and pulse length are to be taken carefully. Likewise, the selection of an appropriate burst-mode level contributes to the efficient partition of the pulse energy into the bursts, as the F/F_{th} relationship showed in figure 114.

6.2 Framework for an efficient laser beam ablation

The idea of this subchapter is to consolidate the actions of this work in the shape of a general framework that includes the methods and their components for a simple understanding and a further efficient application with other types of lasers or materials. The framework includes the experimental approach, with the identification of the most influential parameters, and the analytical approach, with the data collection, model and experimental validation. At the end, the evaluation and analysis of the results and conclusions, represent the application scope of the work.

Figure 116 presents a general framework for an efficient laser beam ablation, which results from all the knowledge, experience and data obtained in this study.

The framework for the analysis and application of an efficient laser beam ablation is formed by two main blocks. The first block corresponds to the experimental approach developed in this work and constituted by eight steps mainly based on quality tools. The definition of the objective of the approach corresponds to the first action of the block, followed by a cause-and-affect study and a strengths-weaknesses-opportunities-threats analysis to gain understanding of the state of the system. After these analyses, preparatory experiments are done to define basic knowledge, such as, the focus position or the number of irradiation layers in the case of laser processing. A core part of the framework in this experimental approach is the identification of influential parameters by the application of the variable search methodology. Once executed the methodology, results are collected and the resulting parameter with significant influence on the objective can be fine-tuned by factorial experiments to gain more information and optimal response to the control parameters. Finally, the laser beam process can be understood from a practical, experimental and application-oriented perspective that might help to optimize resources in an acceptable period of time for industrial fields.

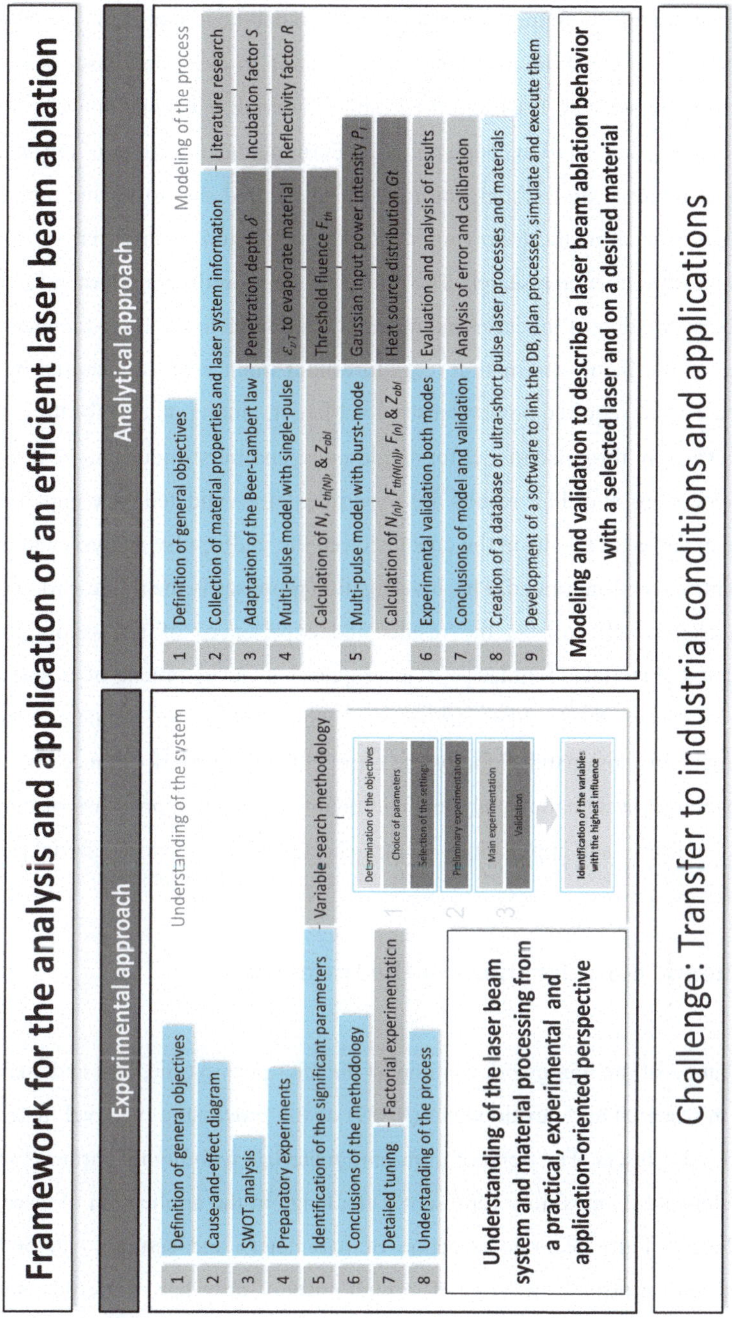

Figure 116: Framework for the analysis and application of an efficient laser beam ablation

The second block of the framework belongs to the analytical approach. The objective of this part is the to achieve the modeling and validation to describe a laser beam ablation behavior with the selected laser and the desired material. It is constituted by nine steps that were explained in detail in chapter 5. After defining the objectives of the analysis, the collection of material and system information is favored due to its importance in the development of each of the following steps. The collection of information is based on literature and experimental research. The following step, similarly important for following steps, is the adaptation of the laser and material conditions to the Beer-Lambert law; this analysis gives the input information (e.g., threshold fluence of the material by the heat transfer theory and required power density to ablate it, among others) for the multi-pulse modeling. Thus, multi-pulse model for single-pulse and burst-mode can be developed in the following steps and their results are experimentally validated. A comparison of the model and experiment results in the conclusion of the multi-pulse analysis and their error can be quantified and calibrated. This investigation proposes in the framework the creation of a database, which collects the results and experiences of ultra-short pulse laser processes and materials. Similarly, it is proposed a development of a software that possesses access to the database and is able to plan processes, simulate behavior and execute processing commands. Finally, the effort of the framework of this work is to target the application of systematic, analytical and quality-oriented approaches of this work to industrial conditions and applications in ultra-short pulse laser beam material processing.

6.3 Economic considerations of the framework

The application of the framework can impact the relationship time, cost and quality of the activities in research and development (R&D) of the processing of novel materials with ultra-short pulse lasers. The trend of ultra-hard materials with higher hardness values and complex element compositions represent a challenge in the application of laser systems. The challenge of the accurate selection of parameters and settings cannot be easily transferred with a prior knowledge. The high costs of ultra-hard materials, such as the combination of tungsten carbide and polycrystalline diamond (e.g., from 50 to 100 euros approximately for a cutting insert blank [146]), increases the importance of reducing destructive experimentation, introducing modeling and executing simulations. Therefore, the framework for the analysis and application of an efficient laser beam ablation offers two kind of solutions for the minimization of experiments and their effects in time and cost. The first approach, the experimental, offers a fast response to the investigation of novel materials and the need of processing information. Figure 117 illustrates schematically the advantages of its use and compares it with a tradition investigation in laser processing.

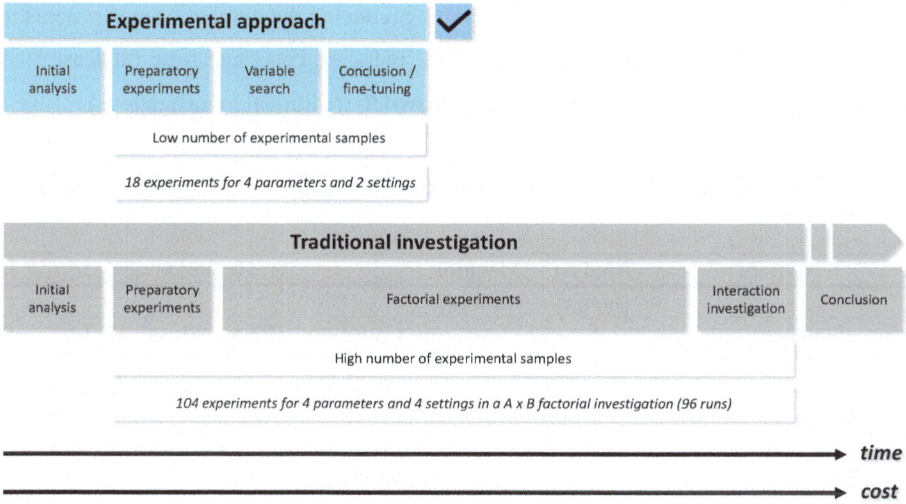

Figure 117: Schematic comparison of the experimental approach with a traditional investigation of laser beam ablation as a function of time and cost

The use of the experimental approach of this work targets the information gain in a shorter period of time and with less experimentation than traditional investigations using factorial experiments and trial/error analysis. The application of the variable search methodology measures the impact of significant parameters and setting avoiding unnecessary experimental runs that might represent higher costs and longer time, e.g., the application of a lower number of experiments: 4 preparatory experiments, 4 preliminary experiments, 8 main experiments and 2 validation experiments, approximately (compared to 4 preparatory experiments, 96 A x B factorial experiments with four parameters and four settings and 4 validation experiments). This advantage makes it attractive for the application in industrial environment working with laser systems, new material and costly resources and applications. Examples of theses application fields are the cutting tool industry, jewelry, time pieces sector and nanotechnology.

The time period breakdown of the activities in the experimental approach of this work is listed in table 31. It presents the time invested by an analyst and the time of machine operation in the application of the approach. The analysis reached the objective of the introduction at the beginning of this work of implementing a quality-oriented approach to achieve improvements for laser beam ablation by reducing the number of experiments in traditional laser beam investigations, and thus, reducing costs and time (at least 50% resources optimization and savings when compared to traditional investigations).

Table 31: Exemplary time breakdown in the application of the experimental approach for laser beam ablation

Activity	Analyst [hours]	Machine [hours]	Remarks
Initial analysis	8	-	Cause-and effect analysis, SWOT, etc.
Preparatory experiments	-	4	Determination of focus position, etc.
Variable search methodology	-	8	Preliminary, main experimentation, validation
Metrology	4	-	Confocal microscopy
Conclusion	4	-	Parameters with significant influence
Fine-tuning	-	4	Degree of influence and parameters interactions
Total	16	16	Estimated 32 hrs of analyst and machine operation
		560 €	Estimated intern machine operation cost of 35 €/hr*

Estimated analyst and machine operation times based on experience in confidential industrial projects
Analysis of one type of material or material with similar physical, chemical and optical properties
Work week period based on 40 working and operation hours, five days a week
* Based on an estimated laser system cost of 600 to 800 k€, 6000 hr/year operation in a 5-year projection and corresponding fixed costs.

The second approach of the framework, the analytical, offers a continuous improvement process that requires of an initial collection of system and material information to develop the described methodology for single-pulse and multi-pulse laser processing. The Beer-Lambert law and multi-pulse analyses offer the adaptability to new materials and different laser conditions. Thus, it is possible to model a laser ablation process for other materials, when the input data is available. For this reason, the creation of a database of ultra-short pulse laser processes and material information accelerates the applicability and extends the possibilities of the framework in every data input. Figure 118 schematizes the scope of the analytical approach.

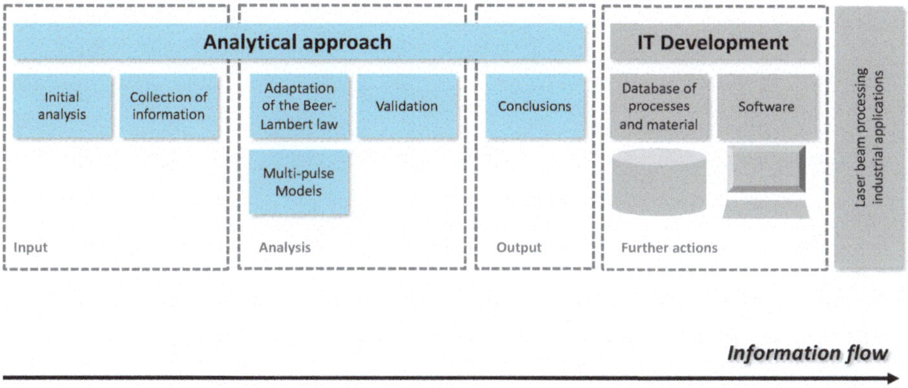

Figure 118: Schematic scope of the analytical approach

The development of a software connected to the database represents an ambitious objective capable of planning processes, simulate them (thanks to models) and execute them through an interface to the operating laser machine. The integration of the theory, experimentation, information management and application of laser beam material processing represents the challenge to direct the scientific and industrial efforts.

7 Conclusion and outlook

The end of the learning cycle of this investigation is located in this chapter; as part of the fourth phase, *act*, of the structure of the thesis, the *conclusion and outlook* aim to summarize the findings of an efficient laser beam ablation on cemented WC and to recommend further actions for the implementation and optimization of the laser beam process in industrial environments. The structure of the thesis, presented in the introduction chapter, illustrates the quality cycle of Deming and the corresponding position of this chapter (figure 4). The phase *act* represents, in a quality-oriented system, the place for supporting, correcting and recommending optimization paths. Therefore, the following lines presents an evaluation of the objectives of this work, summarizes the methodologies and their results and presents specific improvement fields of the laser beam investigation.

This work started with a review of literature and state-of-the-art theories of four relevant areas for this research. The first area was the study of ultra-short pulse lasers to add a deep understanding about fundamentals, working principles, configurations, classifications, and specifically, relevant topics of picosecond lasers. This information contributed with the understanding of ultra-short pulse lasers and thus their application fields and capabilities. The second area corresponded to the understanding of the laser beam ablation process, its definition, thermal and photo-induced considerations and the incubation phenomenon. The third area introduced cemented WC as the choice of material of this work. The fourth area presented quality-oriented tools, e.g., DoE, and variable search methodology, to support the development of a concept for an efficient material laser beam ablation with a picosecond laser in chapter 3. Thus, picosecond laser beam ablation is confirmed as a relevant tool for the ablation of ultra-hard materials and a research concept was developed. Furthermore, two approaches for the investigation are presented from an experimental and analytical perspective, respectively. The methodology for the identification of influential parameters, as the experimental approach in chapter 4, executed a quality-oriented methodology based on the SWOT analysis, cause-and-effect diagram and the variable search methodology. Its application organized a picosecond laser

J. P. Calderón Urbina, *Efficient material laser beam ablation with a picosecond laser*, Light Engineering für die Praxis, https://doi.org/10.1007/978-3-662-61886-8_7

experimental set-up to aim the maximization of the ablation rate on cemented WC without the application of an excessive amount of material samples. The conclusion of the methodology gave the interaction of pulse repetition rate *f* and scanner speed v_s in the form of pulse overlap and track overlap *PO/TO* as the most influential parameter in the maximization of the ablation rate with a value of influence (variation average) of 98%. The second most influential factors resulted in the application of laser beam power P_L with 59% and burst-mode *BM* presented 46%. These parameters have to be observed in that order for the processing of cemented WC in further operations to achieve high values of ablation rate and accurate responses in the picosecond laser beam process.

A core part of this work is presented in chapter 5 as the second approach of the investigation. The description of the model executed a theoretical analysis of the picosecond laser beam ablation of cemented WC by the application of the Beer-Lambert law and multi-pulse ablation modeling. The application of both methodologies required the completion of system and material information that was fulfilled by literature research and specification sheet information. The unavailable material properties were obtained by experimental investigations, like in the cases of the incubation factor and the absorptivity (reflectivity) factor. Based on the theory that the ablation threshold of a material behaves distinctly to different numbers of irradiated laser pulses, the incubation factor was determined by the application of a well-known method that gave a value of incubation factor of $S = 0.75$. Similarly, the search of the reflectivity factor was experimentally executed and showed a value of $R = 0.33$. The completion of the material and system information requirements allowed the application of the adaptation of the Beer-lambert law to picosecond pulse laser ablation. Outstanding results of the adaptation of the Beer-Lambert law are the determination of the threshold fluence for cemented WC thanks to the application of the heat transfer theory and giving a total energy per area to ablate solid cemented WC of $F_{th} = 1960 \text{ J m}^{-2}$. Similarly relevant was the calculation of the input power intensity P_I emitted from the laser, which was adapted to a Gaussian beam profile and based on actual beam measurements of the source. At the end, visualizations of a picosecond laser pulse under the five available pulse repetition rates were modeled. They showed the progression of power density *Gt* as a function of depth *y* in the material and time *t* (pulse length). The depth results were validated by experiments showing satisfying results. The findings from the adaptation of the Beer-Lambert law acted as basis for development of the multi-pulse laser ablation model for both single-pulse mode and burst-mode, respectively. The modeling describes ablation depth Z_{abl} by the definition of the number of pulses N irradiating the same area, the corresponding threshold fluence F_{th} for N and the input fluence F following a Gaussian behavior. The use of burst-mode affects the application of the model by adapting the number of pulses with the corresponding burst-mode value $N_{(n)}$. This fact modifies not only N but also threshold fluence F_{th} (higher number of pulses irradiating the same area and input fluence $F_{(n)}$, partition of pulse energy into bursts). The results of the multi-pulse models with single-pulse and burst-mode were experimentally validated, analyzed and discussed in subchapter 5.4.6. The conclusions allowed to understand the ablation depth behavior of multi-pulse laser beam ablation on cemented WC. For example, the accumulation of energy at low scanner speeds and high pulse repetition rates cause high values of ablation depth. The use of high levels of burst-mode increases the energy accumulation at low scanner speeds and high pulse repetition rates and consequently causes the formation of structures of negative ablation depth. This phenomenon can result attractive for other laser processing applications, such as, laser structuring. A summary of the results of the analytical approach is found in subchapter 5.5 for concise understanding.

Finally, influential parameters and model results are evaluated and analyzed in the last chapter of this work. This part puts forward recommendations from the outcome of the experimental and analytical approach for an efficient laser beam ablation. A second part of the last chapter introduces a framework for an efficient laser beam ablation; this framework systematically concentrates all methods from the experimental and analytical approach for their practical application with new materials and system conditions. It describes schematically the steps to follow for the understanding of the laser beam system and material processing from a practical, experimental and application-oriented perspective (experimental approach); and, the steps for the modeling and validation to describe a laser beam ablation behavior with selected laser and on desired material (analytical approach). In the end, economic considerations for the application of the experimental and analytical approach are presented. It shows a comparison of the experimental approach and traditional investigations, and it describes the advantages in resource optimization when it is applied. Moreover, the work presented a combination of methodologies that accelerates approximately 50% laser beam investigations in novel and compound materials, such as, cemented WC and other ultra-hard materials; and consequently, that reduces in 40% their corresponding costs by the application of approximately 70% less experiments. The use of the analytical approach is described by the efficient and systematic information flow and enhanced by the recommendation of complementing the approach by the integration of an IT development, composed by a database of material and processes and an application software for future incorporation in industrial environments.

Thus, the investigation of laser beam ablation of cemented tungsten carbide by using a picosecond laser was achieved from theoretical analyses and experimental methods. In the same way, the creation of a systematic quality-oriented framework for the analysis and application of an efficient laser beam ablation was presented to guide further efforts on the field.

Further future actions for an efficient laser beam processing of material includes the calibration of the models of this work, the precise analysis of results that are out of the error tolerances and the study of the formation of structures by defining the limits of ablation and the microscopic thermodynamic phenomena. Further improvement fields of this work are the adaptation of the models to volume ablation, a methodology to reduce model error and investigations of the control of pulse energy by changing the levels of laser beam power. A precise analysis and description of the behavior of burst-mode can help gain a more accurate model with this feature. Furthermore, investigations with other trending materials, such as, ultra-hard (e.g., polycrystalline diamond and cubic boron nitride), compound (e.g., carbon fiber reinforced plastic), novel and ceramic materials, are required. Further future actions from a broader perspective include the already mentioned IT development of a database and a laser beam processing software for industrial applications. The latter can allow to fill the database with material and process information from previous experiences, optimize them and execute them in the machine. This effort can evolve into an automated laser beam material processing system that can be upgraded by the integration of new materials. Finally, new quality-oriented tools can be integrated in the methodology to make effective and efficient the experimental efforts in research and industry.

8 Bibliography

[1] Breitling, D., Ruf, A., Dausinger, F. *Fundamental aspects in machining of metals with short and ultrashort laser pulses. In: SPIE Photon Processing in Microelectronics and Photonics III.* 49 - 63. San Jose, Ca. : Proceedings SPIE, 2004. Vol. 5339. doi:10.1117/12.541434.

[2] Dirscherl, M. *Ultrakurzpulslaser, Grundlagen und Anwendungen.* Erlangen : Bayerisches Laserzentrum GmbH, 2005. Bd. II. 3-9809601-2-9.

[3] Gillner, A. *Micro and nano structuring with ultra short pulsed lasers. In: 3rd. International Symposium on Laser Micromachining.* Chemnitz : 3D-Micromac, 2010.

[4] —. *Ultrakurzpulslaser - Das Werkzeug der Zukunft ?! In: 1. Aachener Ultrakurzpuls-Workshop.* [ed.] Fraunhofer ILT. Aachen : s.n., 2011.

[5] Dumitru, G., Romano, V., Weber, H. P., et Al. *Femtosecond ablation of ultrahard materials. In: Applied Physics A – Material, Science & Processing.* [ed.] Springer. 2002. pags. 729 - 739. Vol. 74.

[6] Eberle, G., Wegener, K. *Ablation study of WC and PCD composites using 10 picosecond and 1 nanosecond pulse durations at green and infrared wavelengths. In: 8th International Conference on Photonic Technologies LANE 2014.* Erlangen : Physics Procedia, 2014. pags. 951 – 962. Vol. 56.

[7] Dold, C., Henerichs, M., Gilgen. P., Wegener, K. *Laser processing of coarse grain polycrystalline diamond (PCD) cutting tool inserts using picosecond laser pulses. In: Lasers in Manufacturing Conference 2013.* 603 - 609. Munich : Physics Procedia, 2013. Vol. 41.

[8] Calderón Urbina, J. P., Daniel, C., Emmelmann, C. *Experimental and analytical investigation of cemented tungsten carbide ultra-short pulse laser ablation. In: Lasers in Manufacturing Conference 2013.* Munich : Physics Procedia, 2013. pags. 752 - 758. Vol. 41.

[9] Klimt, B. *Micromachining with ps-laser: Quality, Flexibility, Speed, Cost.* s.l. : Lumera Laser GmbH, 2011.

[10] AB, Sandvik. *Sandvik Products - Sandvik Hyperion: cemented carbide, diamond and CBN.* s.l. : http://www.hyperion.sandvik.com/en/products/, 23.02.2015.

J. P. Calderón Urbina, *Efficient material laser beam ablation with a picosecond laser*, Light Engineering für die Praxis, https://doi.org/10.1007/978-3-662-61886-8

[11] Lasertec, Sauer GmbH. *Mit dem Laser in eine neue Dimension der Schneidkantenbearbeitung. In: http://www.diamond-business.de/index.php/artikel-bibliothek/werkzeug-fertigung/item/mit-dem-laser-in-eine-neue-dimension-der-schneidkantenbearbeitung.* s.l. : Diamond Business, 23.02.2014.

[12] Chase, R. B., Jacobs, F. R., Aquilano, N. J. *Operations management for competitive advantage.* 11th. Boston et Al. : McGraw-Hill Irwin, 2006. 0-07-298-390-6.

[13] Pfeifer, T. *Quality Management; Strategies, Methods, Techniques.* München, Wien : Carl Hanser Verlag, 2002. 3-446-22003-8.

[14] Meijer, J., Du, K., Gillner, A., et al. *Laser Machining by short and ultrashort pulses, state of the art and new opportunities in the age of photons. In: CIRP Annals - Manufacturing technology.* 2. s.l. : Elsevier, 2002. pp. 531-550. Vol. 51.

[15] DIN 32540 : 2012 - 08. *Laser Beam Removing - Thermal removing with laser beam - Definitions, Influence factors, procedures.* Berlin : Beuth, 1997 : 2012.

[16] Meijer, J. *Laser beam machining (LBM), state of the art and new opportunities. In: 14th Interntaional Symposium on Electromachining (ISEM XIV).* 1 - 3. s.l. : Elsevier, 2004. pags. 2 - 17. Vol. 149.

[17] Poprawe, R. *Lasertechnik für die Fertigung - Grundlagen, Perspektiven und Beispiele für den innovativen Ingenieur.* Berlin et Al. : Springer, 2005. 3-540-21406-2.

[18] Kannatey-Asibu, E. Jr. *Principles of laser material processing.* New Jersey : John Wiley & Sons, Inc., 2009. ISBN 978-0-470-17798-3.

[19] Calderon Urbina, J. P., Daniel, C., Emmelmann, C. *Variables search of ultra-short pulse laser ablation of cemented tungsten carbide. Oral presentation. In: LANE 2012 "7th International Conference and Exhibition on photonic technologies".* Fürth : s.n., 2012.

[20] Davim, J. P., Jackson, M. J. *Nano and micromachining.* s.l. : ISTE Ltd and John Wiley & Sons, Inc., 2009. ISBN 978-1-84821-103-2.

[21] Ion, J. C. *Laser processing of engineering materials; principles, procedure and industrial application.* Amsterdam : Elsevier Butterworth-Heinemann, 2005. 978-0080971896.

[22] Steen, W. M., Mazumder, J. *Laser material processing.* 4th Ed. London et Al. : Springer, 2010. ISBN 978-1-84996-061-8.

[23] Stafe, M., Marcu, A. y Puscas, N., N. *Pulsed laser ablation of solids - Basics, theory and applications.* Berlin, Heidelberg : Springer, 2014. Vol. 53. ISBN 978-3-642-40977-6.

[24] Propawe, R. *Tailored light 2 - Laser application technology.* Berlin Heidelberg : Springer, 2011. ISBN 978-3-642-01236-5.

[25] Paschotta, R. Keller, U. *Passively Mode-Locked Solid-State Lasers . In: Solid-state lasers and applications.* [ed.] B. J., Sennaroglu, A. Thompson. Boca Raton : CRC Press, Taylor & Francis Group, 2007. pags. 259-318. ISBN 0-8493-3589-2.

[26] Eichhorn, M. *Laser physics, from principles to practical work in the lab.* Saint Louis, France : Springer, 2014. ISBN 978-3-319-05127-7.

[27] Svelto, O. *Passive mode-locking using a saturable absorber - Principles of Lasers. In: Kannatey-Asibu, E. Principles of laser materials processing.* New Jersey : John Wiley & Sons, Inc., 2009.

[28] Castillejo, M., Ossi, P. M., Zhigilei, L. *Lasers in materials science.* Madrid : Springer, 2014. Vol. 191. ISBN 978-3-319-02897-2.

[29] Keller, U. *Ultrafast solid-state lasers. In: Landolt-Börnstein, Group VIII/1B1, Laser Physics and Applications.* [ed.] G., Weber, H., Proprawe, R. G. Herziger. Berlin : Springer, 2007. pags. 33-167. Vol. 1 Subvolume B. 978-3-540-26033-2.

[30] Strickland, D., Mourou, G. *Compression of amplified chirped optical pulses. In: Optics Communications.* Rochester : Elsevier Science Publishers B.V., 1985. pags. 219-221. Vol. 56.

[31] Kryukov, P. G. *Ultrashort-pulse lasers. In: Quantum Electronics.* Moscow : Kvantovaya Elektronika and Turpion Ltd, 2001. pags. 95-119. Vol. 31.

[32] Hellwarth, R. W. *Control of Fluorescent Pulsations. In: Advances in Quantum Electronics.* [ed.] J. R. Singer. s.l. : Columbia University Press, 1961. pags. 334-341.

[33] Hellwarth, R. W., McClung, F. J. *Giant pulsations from ruby. In: Applied Physics.* 1962. pags. 838-841. Vol. 33.

[34] Zhou, S. *Ultrashort laser pulse generation and amplification in fiber systems, and temporal pulse shaping. Doctoral dissertation at Cornell University.* Ithaca : Cornell University, 2007.

[35] Wolf, E. *Progress in Optics.* Rochester : Elsevier B. V., 2004. Vol. 46. 0444514686.

[36] Paschotta, R. *Field Guide to Laser Pulse Generation.* Bellingham : SPIE Press, 2008. 9780819472489.

[37] Larsen, C., Giesberts, M., et al. *Gain-switched all-fiber laser with narrow bandwidth. In: Optics Express.* [ed.] A. M. Weiner. Lyngby : OSA The Optical Society, 2013. pags. 12302-12308. Vol. 21. doi: 10.1364/OE.21.012302.

[38] Coherent, Inc. *Nanosecond Lasers. https://www.coherent.com/products/?2183/Nanosecond-Lasers.* Santa Clara, Dieburg : Coherent, Inc., Last access: 1.06.2016.

[39] TRUMPF GmbH + Co. KG. *TruMicro Series 7000. Short and ultra-short pulsed lasers. http://www.trumpf-laser.com/en/products/solid-state-lasers/short-and-ultra-short-pulsed-lasers.html.* Ditzingen : TRUMPF GmbH + Co. KG, Last access: 1.06.2016.

[40] ROFIN-SINAR Laser GmbH. *PowerLine SL. Short Pulse Lasers. https://www.rofin.com/en/products/short_pulse_lasers/powerline_sl/.* Hamburg : ROFIN-SINAR Laser GmbH, Last access: 1.06.2016.

[41] JENOPTIK AG. *Nanosecond fiber laser JenLas® fiber ns 20 – 70. Solid State Lasers. Laser and Laser Technology. https://www.jenoptik.com/products/lasers/solid-state-laser-dpss/selector/nanosecond-fiber-laser-jenlas-fiber-ns.* Jena : JENOPTIK AG, Last access: 1.06.2016.

[42] Bengtsson, M., Müller, D., Klimt, B. *Picosecond lasers come of age for micromachining. In: Industrial Laser Solutions for Manufacturing. http://www.industrial-lasers.com/articles/print/volume-28/issue-3/features/picosecond-lasers-come-of-age-for-micromachining.html.* Tulsa : PennWell Corporation, 2013. Last access: 1.06.2016.

[43] Gabdzyl, J. *Application versatility of nanosecond pulsed fiber lasers. In: Industrial Laser Solutions for Manufacturing. http://www.industrial-lasers.com/articles/print/volume-29/issue-2/features/application-versatility-of-nanosecond-pulsed-fiber-lasers.html.* Tulsa : PennWell Corporation, 2014. Last access:1.06.2016.

[44] Eiselen, S., Wu, D., Galarneau, P. et al. *The role of temporal energy input in laser micro machining using nanosecond pulses. In: Lasers in Manufacturing Conference 2013.* Munich : Elsevier - Physics Procedia, 2013. pags. 676-681. Vol. 41.

[45] Chichkov, B.N., Momma, C., Nolte, S., et al. *Femtosecond, picosecond and nanosecond laser ablation of solids. In: Applied Physics A.* Hannover : Springer-Verlag, 1996. pags. 109-115. Vol. 63.

[46] Ippen, E. P., Shank, C. V., Dienes, A. *Passive mode locking of the cw dye laser. In: Applied Physics Letters.* New Jersey : The American Institute of Physics, 1972. pags. 348-350. Vol. 21.

[47] New, G. H. C. *Mode-locking of quasi-continuous lasers. In: Optics Communications.* Belfast : Elsevier Science B.V., 1972. pags. 188-192. Vol. 6.

[48] OSA The Optical Society. *Laser History Timeline. http://www.osa.org/en-us/history/exhibits/laser_history_timeline/.* Washington D.C. : The Optical Society, Last access: 1.06.2016.

[49] Bloembergen, N. *From Nanosecond to Femtosecond Science. In: More Things in Heaven and Earth - A Celebration of Physics at the Millennium.* New York : Springer, 1999. pags. 474-481. 978-1-4612-1512-7.

[50] Shank, C. V., Ippen, E. P. *Subpicosecond kilowatt pulses from a mode-locked cw dye laser. In: Applied Physics Letters.* Holmdel : American Institute of Physics, 1974. pags. 373 - 375. Vol. 24.

[51] Fork, R. L., Greene, B. I. and Shank, C. V. *Generation of optical pulses shorter than 0.1 psec by colliding pulse mode locking. In: Applied Physiscs Letters.* Holmdel : American Institute of Physics, 1981. pags. 671-672. Vol. 38.

[52] Rose, M. *A History Of The Laser: A Trip Through The Light Fantastic. In: Photonics Sprectra. http://www.photonics.com/Article.aspx?AID=42279.* s.l. : Photonics Media, 2010. Last access: 1.06.2016.

[53] Moulton, P. F. *Spectroscopic and laser characteristics of Ti:Al2O3. In: Journal of the Optical Society of America B.* Concord : Optical Society of America, 1986. pags. 125-133. Vol. 3. 10.1364/JOSAB.3.000125.

[54] Fork, R. L., Brito Cruz, C. H., Becker, P. C., et Al. *Compression of optical pulses to six femtoseconds by using cubic phase compensation. In: Optics Letters.* Holmdel : Optical Society of America, 1987. pags. 483-485. Vol. 12. DOI: 10.1364/OL.12.000483.

[55] Blow, K. J., Wood, D. *Mode-locked lasers with nonlinear external cavities. In: Journal of the Optical Society of America B.* Martlesham Heath : Optical Society of America, 1988. pags. 629-632. Vol. 5.

[56] Spence, D. E., Kean, P., N., Sibbett, W. *60-fsec pulse generation from a self-mode-locked Ti:sapphire laser. In: Optics Letters.* s.l. : Optical Society of America, 1991. pags. 42-44. Vol. 16. doi: 10.1364/OL.16.000042.

[57] Kärtner, F. X., Ippen, E. P., and Cundiff, S. T. *Femtosecond laser development. In: Femtosecond optical frequency comb: Principle, operation and applications.* [ed.] J., Cundiff, S. T. Ye. New York : Springer, 2005. pags. 54-77. 978-0-387-23791-6.

[58] Cerullo, G. Manzoni, C. *Solid-State Ultrafast Optical Parametric Amplifiers. In: Solid-state lasers and applications.* [ed.] B. J., Sennaroglu, A. Thompson. Boca Raton : CRC Press, Taylor & Francis Group, 2007. pags. 437-472. 0-8493-3589-2 .

[59] Diels, J.-C., Rudolph, W. *Ultrashort laser pulse phenomena - Fundamentals, techniques, and applications on a femtosecond time scale.* 2nd. Albuquerque : Elsevier Inc., 2006. ISBN: 978-0-12-215493-5.

[60] Keller, U., Miller, D. A. B., Boyd, G. D., et al. *Solid-state low-loss intracavity saturable absorber for Nd:YLF lasers: an antiresonant semiconductor Fabry–Perot saturable absorber.* Holmdel : Optical Society of America, 1992. pags. 505-507. Vol. 17. doi: 10.1364/OL.17.000505.

[61] Keller, U. *"Discovering" the SESAM. In: IEEE Leos. http://photonicssociety.org/newsletters/aug07/SESAM.html.* s.l. : IEEE, 2007. Last accessed: 1.06.2016.

[62] Keller, U., Gallmann, L. *Passive mode-locking. In: ultrafast laser physics. Lecture ETH. https://www.ethz.ch/content/dam/ethz/special-interest/phys/quantum-electronics/ultrafast-laser-physics-dam/education/lectures/ultrafast_laser_physics/lecture_notes/1_Introduction_HS15.pdf.* Zürich : ETH Zürich, 2015. Last accessed: 1.06.2016.

[63] Clark, B., Gunaratne, T. *Interview about Ultrashort pulse laser micromachining: History and future opportunities. In: Industrial Laser Solutions for Manufacturing. http://www.industrial-lasers.com/articles/2014/04/ultrashort-pulse-laser-micromachining-history-and-future-opportunities.html.* Tulsa : PennWell Corporation, 2014. Last access: 1.06.2016.

[64] Lépine, F., Sansone, G., Vrakking, M. J. J. *Molecular applications of attosecond laser pulses. In: Chemical Physics Letters.* Lyon : Elsevier B.V., 2013. pags. 1-14. Vol. 578.

[65] Hentschel, M., Kienberger, R., Spielmann, Ch., et al. *Attosecond metrology. In: Nature.* Wien : Macmillan Magazines Ltd, 2001. pags. 509-513. Vol. 414.

[66] Corkum, P. *Attosecond pulses at last. In Nature.* Ottawa : Macmillan Magazines Ltd, 2000. pags. 845-846. Vol. 403.

[67] Papadogiannis, N. A., Witzel, B., Kalpouzos, C. et al. *Observation of attosecond light localization in higher order harmonic generation. In: Physical Review Letters.* Crete : American Physical Society, 1999. pags. 4289-4292. Vol. 83. DOI:http://dx.doi.org/10.1103/PhysRevLett.83.4289.

[68] Naeser, T. *Attosecond Physics becomes a Milestone. http://www.mpq.mpg.de/4857719/10_05_17).* Garching : Max Planck Institute of Quantum Optics, 2010. Last access: 1.06.2016.

[69] Baltuška, A., Udem, Th., Uiberacker, M., et al. *Attosecond control of electronic processes by intense light fields. In: Nature.* Wien : Nature Publishing Group, 2003. pags. 611-615. Vol. 422.

[70] Goulielmakis, E., Schultze, M., Hofstetter, M. et al. *Single-Cycle Nonlinear Optics. In: Science.* Garching : American Association for the Advancement of Science, 2008. pags. 1614-1617. Vol. 320. DOI: 10.1126/science.1157846.

[71] Laser Focus World. *German researchers achieve new record in ultrafast metrology: 80 attosecond pulses. In Laser Focus World. http://www.laserfocusworld.com/articles/2008/06/german-researchers-achieve-new-record-in-ultrafast-metrology-80-attosecond-pulses.html.* Tulsa : PennWell Corporation, 2008. Last access: 1.06.2016.

[72] Zhao, K., Zhang, Q., Chini, M. et al. *Tailoring a 67 attosecond pulse through advantageous phase-mismatch. In: Optics Letters.* Orlando : Optical Society, 2012. pags. 3891-3893. Vol. 37.

[73] Pfeiffer, A. N. *Attosecond laser systems and applications. In: 8th Annual Laser Safety Officer Workshop.* Berkeley : s.n., 2012.

[74] Ndabashimiye, G., Ghimire, S., Wu, M., et al. *Solid-state harmonics beyond the atomic limit. In: Nature.* Stanford : Macmillan Magazines Ltd, 2016. pags. 520-523. Vol. 534. doi:10.1038/nature17660.

[75] Phys.org. *Creating attosecond laser pulses by inducing 'high harmonic generation' in a solid. In: Phys.org. http://phys.org/news/2016-06-attosecond-laser-pulses-high-harmonic.html.* s.l. : Phys.org, 2016.

[76] Diveki, Z. *Generation and application of attosecond pulses. Ph. D. Thesis.* Paris : Université Paris Sud - Paris XI, 2012.

[77] Chang, Z. *Fundamentals of Attosecond Optics.* Orlando : CRC Press, 2011.

[78] Keller, U. *Ultrafast laser Physics. Vortrag ETH Zürich.* Zürich : ETH Zürich, 2015.

[79] Schafer, K., Gaarde, M. *Attosecond pulse generation & application: Theoretical perspectives from strong field physics. Vortrag Louisiana State University.* Baton Rouge : Louisiana State University, 2006.

[80] Chini, M., Zhao, K., Chang, Z. *The generation, characterization and applications of broadband isolated attosecond pulses. In: Nature Photonics.* Orlando : Macmillan Publishers Limited, 2014. pags. 178-186. Vol. 8. DOI: 10.1038/NPHOTON.2013.362.

[81] Marciak-Kozlowska, J., Kozlowski, M. *On the hyperbolic Pennes equation description of the heat pulse-human cornea interaction. In: Nova Journal of Eye Research.* Warsaw : Nova Science Publishers, 2010. pags. 49-62. Vol. 1.

[82] Hargrove, L. E., Fork, R. L., Pollack, M. A. *Locking of He–Ne laser modes induced by synchronous intracavity modulation. In: Applied Physics Letters.* New Jersey : American Institute of Physics, 1964. pags. 4-5. Vol. 5.

[83] Mocker, H. W., Collins, R. J. *Mode competition and self-locking effects in a Q-switched ruby laser. In: Applied Physics Letters.* s.l. : The American Institute of Physic, 1965. pags. 270-273. Vol. 5.

[84] DeMaria, A. J., Stetser, D. A., and Heynau, H. *Self mode-locking of lasers with saturable absorbers. In: Applied Physics Letters.* s.l. : The American Institute of Physics, 1966. pags. 174-176. Vol. 8.

[85] Shapiro, S. L. *Introduction - A historical overview. In: Topics in applied physics - Ultrashort Light Pulses: Picosecond Techniques and Applications.* Los Alamos : Springer, 1977. Vol. 18.

[86] Hönninger, C. *Industrial Ultrafast Lasers and Applications. In: International Laser and Coating Symposium ILACOS 2012.* Dresden : Amplitude Systèmes, 2012.

[87] Coherent, Inc. *Hyper Rapid. Family of picosecond lasers with 25W, 50W, and 100W at 1064 nm and 1000 kHz. https://www.coherent.com/products/?2075/Hyper-Rapid.* Santa Clara, Dieburg : Coherent, Inc., 2016. Last access: 1.06.2016.

[88] Emmelmann, C., Calderón Urbina, J. P. *Analysis of the Influence of Burst-Mode Laser Ablation by Modern Quality Tools. In: Laser in Manufacturing LiM 2011 Physics Procedia.* Munich : Elsevier, 2011. pags. 172-181. Vol. 12.

[89] Müller, D., Haloui, H., Klimt, B. et al. *Burst-Mode Micromachining with High Power PS Lasers. In: 3rd International Symposium on Laser Micromachining ISL 2010.* Chemnitz : ISL, 3D-Micromac, 2010.

[90] Dold, C. A. *Picosecond laser processing of diamond cutting edges. Doctoral dissertation at ETH Zürich.* Zürich : ETH Zürich, 2013.

[91] Hu, W., Shin, Y. C., King, G. *Modeling of multi-burst mode pico-second laser ablation for improved material removal rate. In: Applied Physics A.* West Lafayette : Springer-Verlag, 2010. pags. 407-414. Vol. 98. DOI: 10.1007/s00339-009-5405-x.

[92] Neuenschwander, B., Kramer, t., Lauer, B et al. *Burst mode with ps- and fs-pulses: Influence on the removal rate, surface quality, and heat accumulation. In: SPIE Proceedings.* San Francisco : SPIE, 2015. Vol. 9350.

[93] Herrmann, T. *Pikosekunden-Laser: Strahlquellen für die moderne Fertigung. Leistungsstarke Systeme eröffnen neue Perspektiven. In: Laser Technik Journal LTJ* . Kaiserlautern : WILEY-VCH Verlag GmbH & Co. KGaA, 2008.

[94] Kulik, C., Ostendorf, A. *Short and ultrashort laser pulses: application driven comparison of sources types. In: 5th International Symposium on Laser Precision Microfabrication.* Bellingham : SPIE Proceeedings, 2004. pags. 689-694. Vol. 5662.

[95] Siegel, F. *Abtragen metallischer Werkstoffe mit Pikosekunden-Laserpulsen für Anwendungen in der Strömungsmechanik. Doctoral Dissertation an der Gottfried Wilhelm Leibniz Universität Hannover.* [ed.] Dr. Dietmar Kracht. Hannover : Laser Zentrum Hannover e.V. LZH, 2011. 978-3-943104-12-7.

[96] Ruf, A. *Modellierung des Perkussionsbohrens von Metallen mit kurz- und ultrakurz-gepulsten Laser. Doctoral Dissertation an der Universität Stuttgart, Forschungsberichte des IFSW* . Stuutgart : Herbert Utz Verlag, 2004. 3-8316-0372-3.

[97] Fischer, T. *Grinding differend – Cutting tools manufacturing with picosecond laser. In: Light 2012.* Hamburg : EWAG Körber AG, 2012.

[98] Wegener, K. Dold, C., Henerichs, M., et al. *Laser Prepared Cutting Tools. In: LANE 2012.* Fürth : Elsevier B. V., Physics Procedia, 2012. pags. 240-248. Vol. 39. doi:10.1016/j.phpro.2012.10.035.

[99] Darif, M. and Semmar, N. *Numerical Simulation of Si Nanosecond Laser Annealing by COMSOL Multiphysics. In: COMSOL Conference.* Hannover : COMSOL, 2008.

[100] Jaeggi, B., Neuenschwander, B., Schmid, M., et al. *Influence of the pulse duration in the ps-regime on the ablation efficiency of metals. In: Lasers in Manufacturing LiM 2011.* Burgdorf : Elsevier, 2011. pags. 164-171. Vol. 12.

[101] Mannion, P. T., Magee, J., Coyne, E., et al. *The effect of damage accumulation behavior on ablation thresholds and damage morphology in ultrafast laser micro-machining of common metals in air. In: Applied Surface Science.* Galway : Elsevier, 2004. pags. 275-287. Vol. 233.

[102] Jee, Y., Becker, M.F. and Walser, R.M. *Laser-induced damage on single-crystal metal surfaces. In: Journal of the Optical Society of America.* Austin : Optical Society of America, 1988. pags. 648-659. Vol. 5.

[103] Jee, Y. *Multiple pulse laser induced damage of single crystal metal surfaces under 1.06 µm pulsed laser irradiation. Doctoral dissertation at University of Texas.* Austin : University of Texas, 1987.

[104] Xiao, S., Gurevich, L., Ostendorf, A. *Incubation effect and its influence on laser patterning of ITO thin film. In: Applied Physics A, Materials Science and Processing.* Bochum : Springer, 2012. pags. 333-338. Vol. 107. DOI 10.1007/s00339-012-6820-y.

[105] Liang, F., Valle, R., Gingras, D. et Al. *Role of ablation and incubation processes on surface nanograting formation. In: Ooptical Materials Express.* s.l. : Optical Society of America, 2011. Vol. 1.

[106] Raciukaitis, G., Brikas, M., Gecys, P., et al. *Accumulation effects in laser ablation of metals with high-repetition-rate lasers.* Vilnius, Lithuania : SPIE Digital Library, 2008. Vol. 7005, 70052L. DOI 10.1117/12.782937.

[107] Jeitschko, W., Pottgen, R., Hoffmann, R.-D. *Structural Chemistry of Hard Materials. In: Riedel, R.: Handbook of ceramics hard materials.* Weinheim : WILEY-VCH Verlag GmbH, 2000.

[108] White, W. B. *Basic Science of Advanced Ceramics. In: Sōmiya, et Al.: Handbook of advanced ceramics.* s.l. : Elsevier Inc., 2003.

[109] Sherman, D., Brandon, D. *Mechanical Properties and their Relation to Microstructure. In: Riedel, R,.: Handbook of ceramic hard materials.* Weinheim : WILEY-VCH Verlag GmbH, 2000.

[110] Upadhyaya, G. S. *Cemented Tungsten Carbides: Production, Properties, and Testing.* s.l. : William Andrew Inc., 1998.

[111] Fukuhara, M. *Cutting Tools. In: Sōmiya, S., et Al.: Handbook of advanced ceramics.* Yokohama : Elsevier Inc., 2003.

[112] Bailey, M. W. *Superabrasive tools: A Brief Introduction. In: Riedel, R.: Handbook of cereamic hard materials.* Weinheim : WILEY-VCH Verlag GmbH, 2000.

[113] Lengauer, W. *Transition Metal Carbides, Nitrides, and Carbonitrides. In: Riedel. R.: Handbook of Ceramic Hard Materials.* Weinheim : WILEY-VCH Verlag GmbH, 2000.

[114] Cardarelli, F. *Materials Handbook: A Concise Desktop Reference.* Tucson : Springer, 2008.

[115] Bailey, M. W. *Industrial Applications of Diamond and cBN. In: Riedel, R.: Handbook of cereamic hard materials.* Weinheim : WILEY-VCH Verlag GmbH, 2000.

[116] Riedel, R. *Handbook of ceramic hard materials.* Darmstadt : Wiley-VCH, 2000.

[117] Sōmiya, S., Claussen, N., Uchino, K., et Al. *Handbook of Advanced Ceramics.* Tokyo : Elsevier Ltd., 2003.

[118] Materials, AZO. *Tungsten carbide – An overview. https://www.azom.com/properties.aspx?ArticleID=1203; 15.02.2019.* 2019.

[119] Schlottmann, U. *SIDS Initial Assessment Report for SIAM 21 – Tungsten Carbide.* Washington D.C. : undesministerium fuer Umwelt, Naturschutz und Reaktorsicherheit, 2005.

[120] Harrison, P. M., Henry, M., Brownell, M. *Laser processing of polycristaline diamond, tungsten carbide and a related composite material. In: Journal of laser applications.* Crawley : s.n., 2006. pags. 177-187. Vol. 18.

[121] Harrison, P. M., Henry, M., Henderson, I., et Al. *Laser milling of metallic and non-metallic substrates in the nanosecond regime with Q- switched diode pumped solid state lasers. In: Proceedings of SPIE,.* Crawley : s.n., 2004. pags. 624-633. Vol. 5448.

[122] Back, S., Weigel, H. *Design for Six Sigma. Kompaktes Wissen, Konkrete Umsetzung, Praktische Arbeitshilfen.* München : Carl Hanser Verlag, 2014. ISBN 978-3-446-44046-3.

[123] Taghizadegan, S. *Essentials of Lean Six Sigma.* 1. Burlington : Elsevier Inc., 2006. ISBN: 978-0-12-370502-0.

[124] Kleppmann, W. *Versuchsplanung. Produkte und Prozesse optimieren.* 7. München et al. : Hanser, 2011. E-Book-ISBN 978-3-446-42942-0 .

[125] Beckford, J. *Quality: A Critical Introduction.* 2. New York : Routledge, 2002.

[126] Hering, E., Triemel, J., Blank, H.-P. *Qualitätsmanagement für Ingenieure.* 5. Berlin : Springer, 2003.

[127] Bhote, K. R., Bhote, A. K. *World class quality – Using design of experiments to make it happen.* New York : Amacom, 2000.

[128] Lorenz, A., Rehme, O., Emmelmann, C. *Statistiche Versuchsplanung in der Lasermaterialbearbeitung. In: Photonik.* 2003. pags. 93-95. Vol. 3.

[129] Lumera Laser GmbH - Coherent, Inc. *HyperRapid 50 - High Power Industrial ps-Laser - Datasheet.* Kaiserlautern / Santa Clara, California : s.n., 2010 / 2018.

[130] Keyence Corporation. *VK-X Series; Automatically Measure and Analyze 3D Surfaces. https://www.keyence.com/landing/microscope/lp_vkx.jsp.* s.l. : Keyence Corporation, Last access: 1.09.2017.

[131] Emmelmann, C., Calderon Urbina, J. P. *Process design of ultra-short pulse laser ablation by modern quality-oriented design of experiments. In: International Conference on Innovative Technologies, IN-TECH 2012.* Rijeka : s.n., 2012.

[132] CVI Melles Griot. *Basic laser principles. In: Introduction to Laser Technology. http://www.astro.caltech.edu/~lah/ay105/pdf/Laser-Guide.pdf.* Rochester : CVI Melles Griot, 2009 Last access: 26.07.2017. Vol. II.

[133] Förster, D. J., Bui, D. A., Onuseit, V., et Al. *Energy transfer mechanisms during laser pulsed processing of metals. In: Lasers in Manufacturing Conference LiM 2015.* München : WTL Wissentschaftliche Gesellschaft Lasertechnik e.V., 2015.

[134] Liu, J. M. *Simple technique for the measurements of pulsed Gaussian-beam spot sizes. In: Optics Letters.* Cambridge : Optical Society of America, 1982. pags. 196-198. Vol. 7. DOI 0146-9592/82/050196-03.

[135] Di Niso, F., Gaudiuso, C., Sibillano, T., et Al. *Influence of the repetition rate and pulse duration on the incubation effect in multiple-shots ultrafast laser ablation of steel. In: Lasers in Manufacturing Conference LiM 2013.* Munich : Elsevier B.V., 2013. pags. 691-700. Vol. 41. Physics Procedia. DOI 10.1016/j.phpro.2013.03.136.

[136] Di Niso, F., Gaudiuso, C., Sibillano, T., et al. *Role of heat accumulation on the incubation effect in multi-shot laser ablation of stainless steel at high repetition rates.* Bari, Italy : The Optical Society OSA, 2014. Vol. 22. DOI 10.1364/OE.22.012200.

[137] Byskov-Nielsen, J., Savolainen, J.-M., Christensen, M. S., et al. *Ultra-short pulse laser ablation of metals: threshold fluence, incubation coefficient and ablation rates. In: Applied Physics A, Materials Science & Processing.* Aarhus : Springer, 2010. pags. 97-101. Vol. 101. DOI 10.1007/s00339-010-5766-1.

[138] De la Perelle, E. T., Moss T. S., Herbert H. *The measurements of absorptivity and reflectivity. In: Infrared Physics.* Farnborough : Pergamon Press Ltd., 1963. pags. 35-43. Vol. 3.

[139] Zwinkels, J. C., Noël, M. and Dodd C. X. *Procedures and standards for accurate spectrophotometric measurements of specular reflectance. In: Applied Optics.* Ottawa : The Optical Society, 1994. pags. 7933-7944. Vol. 33. https://doi.org/10.1364/AO.33.007933.

[140] Willis, D. A., Xu, X. *Heat transfer and phase change during picosecond laser ablation of nickel. In: International Journal of Heat and Mass Transfer.* West Lafayette : Elsevier, 2002. pags. 3911–3918. Vol. 45.

[141] Reger, D. L., Goode, S. R, Ball, D. W. *Chemistry: Principles and practice.* 3rd. Belmont : Cengage Learning, 2009.

[142] Momma, C., Chichkov, B. N., Nolte, S., et al. *Short-pulse laser ablation of solid targets. In: Optics Communications.* Hannover : Elsevier, 1996. pags. 134-142. Vol. 129. PII SOO30-4018(96)00250-7.

[143] Ströbel, J., Voisat, B., Lagsani, A. F. *Innovative and multifunctional system for homogeneous micrometer structures - MIAMI. In: 7th International Summer School on "Trends and new developments in Laser Technology".* Dresden : s.n., 2018.

[144] König, J., Nolte, S. and Tünnermann, A. *Plasma evolution during metal ablation with ultrashort laser pulses. In: Optics Express 10602.* Stuttgart : Optical Society of America, 2005. Vol. 13.

[145] Kuntze, T. *The importance of a proper process monitoring in (laser) process development. In: 7th International Summer School "Trends and new developments in Laser Technology 2018".* Dresden : Fraunhofer IWS, 2018.

[146] Hyperion. *Hyperion Materials and technologies. Toolmaker Solutions. https://www.hyperionmt.com/en/.* s.l. : Hyperion, 2019. Last access: 1.04.2019.

[147] Khanin, Y. I. *Fundamentals of laser dynamics.* Cambridge : Cambridge International Science Publishi, 2006. 1904602118.

The following student works were done within the scope of the present doctoral thesis:

- Irshad, F.: *Analysis and optimization of material removal using picosecond lasers*. Project work. TU Hamburg, Hamburg, 2010
- Irshad, F.: *Analysis of Surface Structuring using a Picosecond Laser*. Master thesis. TU Hamburg, Hamburg, 2011
- Patel, C. M.: *Analytical investigation of ultra-short pulse laser*. Project work. TU Hamburg, Hamburg, 2012
- González Miramón, S.: *Economic analysis of ultra-short pulse laser ablation*. Master thesis. TU Hamburg, Hamburg, 2012
- Portilla Vazquez, F.: *Analysis and optimization of experimental cutting insert production by picosecond laser beam ablation using design of six sigma methodology*. Master thesis. TU Hamburg, Hamburg, 2012
- Bezborooa, D.: *Literature review: Investigation of picosecond laser sources*. Project work. TU Hamburg, Hamburg, 2013

The manufacturer's authorised representative in the EU is Springer
Nature Customer Service Centre GmbH, Europaplatz 3, 69115 Heidelberg,
Germany. If you have any concerns regarding our products, please
contact ProductSafety@springernature.com

Printed and bound by CPI Group (UK) Ltd, Croydon, CR0 4YY
30/00/2026
02225434-0019